The SCIENTIFIC REVOLUTION

science·culture

A series edited by Steven Shapin

The SCIENTIFIC REVOLUTION

STEVEN SHAPIN

THE UNIVERSITY OF CHICAGO PRESS
CHICAGO AND LONDON

The University of Chicago Press, 60637
The University of Chicago Press, Ltd., London

Paperback edition 1998
Printed in the United States of America
05 04 03 02 01 00 99 98 3 4 5
ISBN: 0-226-75020-5 (cloth)
0-226-75021-3 (paper)

Library of Congress Cataloging-in-Publication Data

Shapin, Steven.
The scientific revolution / Steven Shapin
p. cm.
Includes bibliographical references and index.
ISBN 0-226-75020-5 (cloth : alk. paper). —
ISBN 0-226-75021-3 (pbk. : alk. paper)
1. Science—History. I. Title.
Q125.S5166 1996
509—dc20 96-13196
CIP

∞The paper used in this publication meets the minimum requirements of the American National Standard for Information Sciences—Permanence of Paper for Printed Library Materials, ANSI Z39.48-1992.

For Abigail

Contents

Illustrations

Photo Credits

I thank the following institutions for permission to publish the illustrations reproduced here: the Bancroft Library, University of California, Berkeley (figs. 1 and 20); the National Museum of American History, Washington, D.C. (fig. 6); the Syndics of Cambridge University Library (figs. 13, 14, 17, 23, and 25); the Burndy Library, Cambridge, Massachusetts (fig. 15); and Edinburgh University Library (figs. 21 and 22).

Acknowledgments

This is a work of critical synthesis, not one of original scholarship. Although its aim is to give an up-to-date interpretation of the Scientific Revolution, taking account of much historical research produced in the past ten or fifteen years, it nevertheless draws on the research of generations of scholars. Accordingly, the greatest debts I wish to acknowledge are to the many other historians whose work I use so freely, and whose books and papers are listed in the accompanying bibliographic essay. There should be no doubt about the legitimate sense in which this is as much their book as mine, yet I must also acknowledge that the interpretations I put on their work, and the way I organize their disparate findings and claims, reflect my own particular point of view. For this, of course, I accept entire responsibility.

To enable this book to address a general readership most effectively, and to make the exposition flow as smoothly as possible, I chose—with some concern about setting aside conventions traditionally observed in works by specialist scholars written for other specialists—not to burden the text with dense citations of relevant secondary literature. Moreover, direct quotations from modern historians were reserved for just a few occasions when I judged that their particular ways of putting things were uniquely effective or revealing, or when their precise formulations had attained something like "proprietary" status.

Since my aim was to write a short text that might be useful for teaching, I have, over many years, tried out various versions of this book's accounts and arguments with my students, especially those at Edinburgh University when, during the 1970s and 1980s, I taught the history of science. Whether explicitly or implicitly, these are the people who told me most effectively whether I was making myself understood and, indeed, whether I was making sense. I thank them all.

I am also fortunate in having a few academic colleagues and friends who repeatedly told me that such a book might be useful, who encouraged me through some particularly troublesome passages in its career, and who read earlier versions, making valuable suggestions about many aspects of its content, organization, and presentation. In this connection it is a pleasure to acknowledge the special contributions of Peter Dear and Simon Schaffer. No one familiar with their work could possibly associate them with this book's remaining faults. Two anonymous readers for the University of Chicago Press wrote constructive and detailed reports far beyond the usual call of duty. For assistance in locating several of the illustrations, I am grateful to Paula Findlen, Karl Hufbauer, Christine Ruggere, Simon Schaffer, and Deborah Warner. I thank Alice Bennett, senior manuscript editor at the University of Chicago Press, whose diligent and dedicated copyediting did much to make my writing clearer and less fussy. My editor, Susan Abrams, has throughout given the support and advice for which she has become so well known and so highly respected.

Introduction

The Scientific Revolution: The History of a Term

There was no such thing as the Scientific Revolution, and this is a book about it. Some time ago, when the academic world offered more certainty and more comforts, historians announced the real existence of a coherent, cataclysmic, and climactic event that fundamentally and irrevocably changed what people knew about the natural world and how they secured proper knowledge of that world. It was the moment at which the world was made modern, it was a Good Thing, and it happened sometime during the period from the late sixteenth to the early eighteenth century. In 1943 the French historian Alexandre Koyré celebrated the conceptual changes at the heart of the Scientific Revolution as "the most profound revolution achieved or suffered by the human mind" since Greek antiquity. It was a revolution so profound that human culture "for centuries did not grasp its bearing or meaning; which, even now, is often misvalued and misunderstood." A few years later the English historian Herbert Butterfield famously judged that the Scientific Revolution "outshines everything since the rise of Christianity and reduces the Renaissance and Reformation to the rank of mere episodes. . . . [It is] the real origin both of the modern world and of the modern mentality." It was, moreover, construed as a conceptual revolution, a fundamental reordering of our ways of *thinking* about the natural. In this respect, a story about the Scientific Revolution might be adequately told through an account of radical changes in the fundamental categories of thought. To Butterfield, the mental changes making up the Scientific Revolution were equivalent to "putting on a new pair of spectacles." And to A. Rupert Hall it was nothing less than "an *a priori* redefinition of the objects of philosophical and scientific inquiry."

This conception of the Scientific Revolution is now encrusted with tradition. Few historical episodes present themselves as more substantial or more self-evidently worthy of study. There is an established place for accounts of the Scientific Revolution in the Western liberal curriculum, and this book is an attempt to fill that space economically and to invite further curiosity about the making of early modern science.[1] Nevertheless, like many twentieth-century "traditions," that contained in the notion of the Scientific Revolution is not nearly as old as we might think. The phrase "the Scientific Revolution" was not in common use before Alexandre Koyré gave it wider currency in 1939. And it was not until 1954 that two books—written from opposite ends of the historiographic spectrum—used it as a main title: A. Rupert Hall's Koyré-influenced *The Scientific Revolution*[2] and a volume of J. D. Bernal's Marxist *Science in History* called *The Scientific and Industrial Revolutions.* Although many seventeenth-century practitioners expressed their intention to bring about radical intellectual change, they used no such term to refer to what they were doing.

1. "Early modern," in historians' usage, generally refers to the period in European history from roughly 1550 to 1800. I shall be using the term in a slightly more restrictive sense, to denote the period ending about 1700–1730. Later I will use the terms "modern" and "modernist" to designate some specific reforms of knowledge and practice set on foot in the seventeenth century.

2. In the 1930s the French philosopher Gaston Bachelard referred to "mutations" (or large-scale discontinuities) in the development of the conceptual structure of science, a usage Koyré soon developed: "The scientific revolution of the seventeenth century was without doubt such a mutation. . . .It was a profound intellectual transformation of which modern physics . . . was both the expression and the fruit."

From antiquity through the early modern period, a "revolution" invoked the idea of a periodically recurring cycle. In Copernicus's new astronomy of the mid-sixteenth century, for example, the planets completed their revolutions round the sun, while references to political revolutions gestured at the notion of ebbs and flows or cycles—fortune's wheel—in human affairs. The idea of revolution as a radical and irreversible reordering developed together with linear, unidirectional conceptions of time. In this newer conception revolution was not recurrence but its reverse, the bringing about of a new state of affairs that the world had never witnessed before and might never witness again. Not only this notion of revolution but also the beginnings of an idea of revolution in science date from the eighteenth-century writings of French Enlightenment philosophes who liked to portray themselves, and their disciplines, as radical subverters of ancien régime culture. (Some of the seventeenth-century writers this book is concerned with saw themselves not as bringing about totally new states of affairs but as restoring or purifying old ones.) The notion of a revolution as epochal and irreversible change, it is possible, was first applied in a systematic way to events in science and only later to political events. In just this sense, the first revolutions may have been scientific, and the "American," "French," and "Russian Revolutions" are its progeny.

As our understanding of science in the seventeenth century has changed in recent years, so historians have become increasingly uneasy with the very idea of "the Scientific Revolution." Even the legitimacy of each word making up that phrase has been individually contested. Many historians are now no longer satisfied that there was any singular and discrete event, localized in time and space, that can be pointed to as "the" Scientific Revolution. Such historians now reject even the notion that there was any single coherent cultural entity called "science" in the seventeenth century to undergo revolutionary change. There was, rather, a diverse array of cultural practices aimed at understanding, explaining, and controlling the natural world, each with different characteristics and each experiencing different modes of change. We are now much more dubious of claims that there is anything like "a scientific method"—a coherent, universal, and efficacious set of procedures for making scientific knowledge—and still more skeptical of stories that locate its origin in the seventeenth century, from which time it has been unproblematically passed on to us. And many historians do not now accept that the changes wrought on scientific beliefs and practices during the seventeenth century were as "revolutionary" as has been widely portrayed. The continuity of seventeenth-century natural philosophy with its medieval past is now routinely asserted, while talk of "delayed" eighteenth- and nineteenth-century revolutions in chemistry and biology followed hard upon historians' identification of "the" original Scientific Revolution.

Why Write about the Scientific Revolution?

There are still other reasons for historians' present uneasiness with the category of the Scientific Revolution as it has been customarily construed. First, historians have in recent years become dissatisfied with the traditional manner of treating ideas as if they floated freely in conceptual space. Although previous accounts framed the Scientific Revolution in terms of autonomous ideas or disembodied mentalities, more recent versions have insisted on the importance of situating ideas in their wider cultural and social context. We now hear more than we used to about the relations between the scientific changes of the seventeenth century and changes in religious, political, and economic patterns. More fundamentally, some historians now wish to understand the concrete human *practices* by which ideas or concepts are made. What did people *do* when they made or confirmed an observation, proved a theorem, performed an experiment? An account of the Scientific Revolution as a history of free-floating concepts is a very different animal from a history of concept-making practices. Finally, historians have become much more interested in the "who" of the Scientific Revolution. What kinds of people wrought such changes? Did everyone believe as they did, or only a very few? And if only a very few took part in these changes, in what sense, if at all, can we speak of the Scientific Revolution as effecting massive changes in how "we" view the world, as the moment when modernity was made, for "us"? The cogency of such questions makes for problems in writing as unreflectively as we used to about the Scientific Revolution. Responding to them means that we need an account of changes in early modern science appropriate for our less confident, but perhaps more intellectually curious, times.

Yet despite these legitimate doubts and uncertainties there remains a sense in which it is possible to write about the Scientific Revolution unapologetically and in good faith. There are two major considerations to bear in mind here. The first is that many key figures in the late sixteenth and seventeenth centuries vigorously expressed *their* view that they were proposing some very new and very important changes in knowledge of natural reality and in the practices by which legitimate knowledge was to be secured, assessed, and communicated. They identified *themselves* as "moderns" set against "ancient" modes of thought and practice. Our sense of radical change afoot comes substantially from them (and those who were the object of their attacks), and is not simply the creation of mid-twentieth-century historians. So we can say that the seventeenth century witnessed some self-conscious and large-scale attempts to change belief, and ways of securing belief, about the natural world. And a book about the Scientific Revolution can legitimately tell a story about those attempts, whether or not they succeeded, whether or not they were contested in the local culture, whether or not they were wholly coherent.

But why do we tell *these* stories instead of others? If different sorts of seventeenth-century people believed different things about the world, how do we assemble our cast of characters and associated beliefs? Some "natural philosophers," for example, advocated rational theorizing, while others pushed a program of relatively atheoretical fact collecting and experimentation.[3] Mathematical physics was,

3. In the seventeenth century the word "science" (from the Latin *scientia,* meaning knowledge or wisdom) tended to designate any body of properly constituted

for example, a very different sort of practice from botany. There were importantly different versions of what it was to do astronomy and believe as an astronomer believed; the relations between the "proper sciences" of astronomy and chemistry and the "pseudosciences" of astrology and alchemy were intensely problematic; and even the category of "nature" as the object of inquiry was understood in radically different ways by different sorts of practitioners. This point cannot be stressed too strongly. The cultural practices subsumed in the category of the Scientific Revolution—however it has been construed—are not coextensive with early modern, or seventeenth-century, science. Historians differ about which practices were "central" to the Scientific Revolution, and participants themselves argued about which practices produced genuine knowledge and which had been fundamentally reformed.

More fundamentally for criteria of selection, it ought to be understood that "most people"—even most educated people—in the seventeenth century did not believe what expert scientific practitioners believed, and the sense in which "people's" thought about the world was revolutionized at that time is very limited. There should be no doubt whatever that one could write a convincing history of seventeenth-century thought about nature without even *mentioning* the Scientific Revolution as traditionally construed.

The very idea of the Scientific Revolution, therefore, is at least partly an expression of "our" interest in our ancestors, where "we" are late twentieth-century scientists and those for whom what they believe counts as truth about the natural world. And this interest provides the second legitimate justification for writing about the Scien-

---

knowledge (that is, knowledge of necessary universal truths), while inquiries into what sorts of things existed in nature and into the causal structure of the natural world were referred to, respectively, as "natural history" and "natural philosophy." In the main, this book will follow early modern usage, including the designation of relevant practitioners as natural philosophers, natural historians, mathematicians, astronomers, chemists, and so forth. The term "scientist" was invented only in the nineteenth century and was not in routine use until the early twentieth.

tific Revolution. Historians of science have now grown used to condemning "present-oriented" history, rightly saying that it often distorts our understanding of what the past was like in its own terms. Yet there is absolutely no reason we should not want to know how we got from there to here, who the ancestors were, and what the lineage is that connects us to the past. In this sense a story about the seventeenth-century Scientific Revolution can be an account of those changes that we think led on—never directly or simply, to be sure—to certain features of the present in which, for certain purposes, we happen to be interested. To do this would be an expression of just the same sort of legitimate historical interest displayed by Darwinian evolutionists telling stories about those branches of the tree of life that led to human beings—without assuming in any way that such stories are adequate accounts of what life was like hundreds of thousands of years ago. There is nothing at all wrong about telling such stories, though one must always be careful not to claim too much scope for them. Stories about the ancestors as ancestors are not likely to be sensitive accounts of how it was in the past: the lives and thoughts of Galileo, Descartes, or Boyle were hardly typical of seventeenth-century Italians, Frenchmen, or Englishmen, and telling stories about them geared solely to their ancestral role in formulating the currently accepted law of free fall, the optics of the rainbow, or the ideal gas law is not likely to capture very much about the meaning and significance of their own careers and projects in the seventeenth century.

The past is not transformed into the "modern world" at any single moment: we should never be surprised to find that seventeenth-century scientific practitioners often had about them as much of the ancient as the modern; their notions had to be successively transformed and redefined by generations of thinkers to become "ours." And finally, the people, the thoughts, and the practices we tell stories about as "ancestors," or as the beginnings of our lineage, always reflect some present-day interest. That we tell stories about Galileo, Boyle, Descartes, and Newton reflects something about our late twentieth-century scientific beliefs and what we value about those beliefs. For different purposes we could trace aspects of the modern world back to philosophers "vanquished" by Galileo, Boyle, Descartes, and Newton, and to views of nature and knowledge very different from those elaborated by our officially sanctioned scientific ancestors. For still other purposes we could make much of the fact that most seventeenth-century people had never heard of our scientific ancestors and probably entertained beliefs about the natural world very different from those of our chosen forebears. Indeed, the overwhelming majority of seventeenth-century people did not live in Europe, did not know that they lived in "the seventeenth century," and were not aware that a Scientific Revolution was happening. The half of the European population that was female was in a position to participate in scientific culture scarcely at all, as was that overwhelming majority—of men and women—who were illiterate or otherwise disqualified from entering the venues of formal learning.

Some Historiographical Issues

I mean this book to be historiographically up to date—drawing on some of the most recent historical, sociological, and philosophical engagements with the Scientific Revolution. On the other hand, I do not mean to trouble readers with repeated references to methodological and conceptual debates among academics. This book is not written for professional specialized scholars, and readers who develop an interest in the academic state of play will find guidance in the accompanying bibliographic essay. There is no reason to deny that this story about the Scientific Revolution represents a particular point of view, and that, although I help myself freely to the work of many distinguished scholars, its point of view is my own. Other specialists will doubtless disagree with my approach—some vehemently—and a large number of existing accounts do offer a quite different perspective on what is worth telling about the Scientific

Revolution. The positions represented here on some recent historiographic issues can be briefly summarized:

1. I *take for granted* that science is a historically situated and social activity and that it is to be understood in relation to the *contexts* in which it occurs. Historians have long argued whether science relates to its historical and social contexts or whether it should be treated in isolation. I shall simply write about seventeenth-century science as if it were a collectively practiced, historically embedded phenomenon, inviting readers to see whether the account is plausible, coherent, and interesting.

2. For a long time, historians' debates over the propriety of a sociological and a historically "contextual" approach to science seemed to divide practitioners between those who drew attention to what were called "intellectual factors"—ideas, concepts, methods, evidence—and those who stressed "social factors"—forms of organization, political and economic influences on science, and social uses or consequences of science. That now seems to many historians, as it does to me, a rather silly demarcation, and I shall not waste readers' time here in reviewing why those disputes figured so largely in past approaches to the history of early modern science. If science is to be understood as historically situated and in its collective aspect (i.e., sociologically), then that understanding should encompass all aspects of science, its ideas and practices no less than its institutional forms and social uses. Anyone who wants to represent science sociologically cannot simply set aside the body of what the relevant practitioners *knew* and how they went about obtaining that knowledge. Rather, the task for the sociologically minded historian is to display knowledge making and knowledge holding *as social processes.*

3. A traditional construal of "social factors" (or what is sociological about science) has focused on considerations taken to be "external" to science proper—for example, the use of metaphors from the economy in the development of scientific knowledge or the ideological uses of science in justifying certain sorts of political arrangements. Much fine historical work has been done based on such a construal. However, the identification of what is sociological about science with

what is external to science appears to me a curious and a limited way of going on. There is as much society inside the scientist's laboratory, and internal to the development of scientific knowledge, as there is outside. And in fact the very distinction between the social and the political, on the one hand, and "scientific truth," on the other, is partly a cultural product of the period this book discusses. What is commonsensically thought of as science in the late twentieth century is in some measure a product of the historical episodes we want to understand here. Far from matter-of-factly treating the distinction between the social and the scientific as a resource in telling a historical story, I mean to make it into a topic of inquiry. How and why did we come to think that such a distinction is a matter *of course*?

4. I do not consider that there is anything like an "essence" of seventeenth-century science or indeed of seventeenth-century reforms in science. Consequently there is no single coherent story that could possibly capture all the aspects of science or its changes in which we late twentieth-century moderns might happen to be interested. I can think of no feature of early modern science that has been traditionally identified as its revolutionary essence that did not have significantly variant contemporary forms or that was not subjected to contemporary criticism by practitioners who have also been accounted revolutionary "moderns." Since in my view there is no essence of the Scientific Revolution, a multiplicity of stories can legitimately be told, each aiming to draw attention to some real feature of that past culture. This means that selection is a necessary feature of *any* historical story, and there can be no such thing as definitive or exhaustive history, however much space the historian takes to write about any passage of the past. What we select inevitably represents our interests, even if we aim all the while to "tell it like it really was." That is to say, there is inevitably something of "us" in the stories we tell about the past. This is the historian's predicament, and it is foolish to think there is some method, however well intentioned, that can extricate us from this predicament.

The interpretations of professional historians respect the vast body of factual knowledge we now have about the past. Such respect rightly counts as a measure of intellectual honesty, and all historians wishing to be honest will feel the desire to make endless qualifications to *any* generalization about past science. It is a pull I feel as strongly as any other historian: in the pages that follow there are many summaries I wish I had space to make more nuanced and more qualified. Yet succumbing to that pull has its costs. Stories of endless complexity, endlessly qualified, hedged about with modifications and surrounded by a moat of literature citations, are unlikely to be read by any but specialists. And though such accounts can further our stock of factual knowledge about the past, they are less likely to be coherent enough to advance our overall understanding. Part of my brief, to be sure, is to draw attention to the cultural heterogeneity of seventeenth-century science, but I have elected to do so by following a relatively small number of issues and themes through the period of interest.

I am content to accept that this account of the Scientific Revolution is selective and partial. There is a moderate bias toward the empirical and experimental sciences and toward English materials. This is partly due to my own historical interests and partly the consequence of my judgment that many previous historical surveys have been excessively skewed toward mathematical physics and Continental settings.[4] This concentration was justified by the view that what was "really new" and "really important" in the seventeenth century was the mathematization of the study of motion and the destruction of the Aristotelian cosmos—hence a tight focus upon such figures as Galileo, Descartes, Huygens, and Newton. The pride of place accorded in some traditional stories to mathematical physics and astronomy has tended to give an impression that these practices solely constituted the Scientific Revolution, or even that an account of them counts as what deserves telling about important novelty in early modern science. In weakened form, there is much about these as-

4. In many cases I use English materials not to imply or assert the centrality of developments particular to England but as a way of locally illustrating tendencies that were, in general form, widely distributed in Europe.

sumptions that is worth retaining, but this book will intermittently draw attention to the significance of reformed practices of making observations and constituting *experience* in a wider range of sciences. Indeed, some recent historical work has claimed that the seventeenth century, and especially the English setting, witnessed remarkable innovations in the modes of identifying, securing, validating, organizing, and communicating experience, and I want this survey to reflect the significance of those claims. Nor, despite the fact that this book devotes much attention to what have been called the "mechanical," the "experimental," and the "corpuscular" philosophies, do I simply equate these practices with the Scientific Revolution. Not all seventeenth-century natural philosophy was mechanical or experimental, and among those versions that did embrace mechanism and experimentation, their proper scope and role were disputed. Nevertheless, I think that attempts to "mechanize" not only nature but the means of knowing about nature, as well as *conflicts* over the propriety of mechanical and experimental modes, do capture quite a lot that is worth understanding about cultural change in this period.

If there is any originality about the conception of this book, it possibly flows from its basic organization. The three chapters deal sequentially with what was known about the natural world, how that knowledge was secured, and what purposes the knowledge served. What, how, and why. Some existing surveys have focused almost exclusively on what, while accounts of how have tended to suffer from idealization and why has scarcely been addressed at all, and then in relative isolation from the what and the how.

I want to engage with and to summarize a more-or-less canonical account of changes in belief widely said to be characteristic of the Scientific Revolution, while giving some indication that relevant beliefs varied and were even strongly contested. I start by picking up a number of strands in changing patterns of belief about nature that have routinely been treated by previous historians. I have claimed that there is no essence of the Scientific Revolution, yet pragmatic criteria push me at times toward an artificially coherent account of distinctive changes in natural knowledge. (When that artificial coherence appears, the most I can do is to signal it and, from time to time, point to problems associated with it.)

I shall be drawing special attention to four interrelated aspects of changes in knowledge about the natural world and changes in means of securing that knowledge. First, the mechanization of nature: the increasing use of mechanical metaphors to construe natural processes and phenomena; second, the depersonalization of natural knowledge: the growing separation between human subjects and the natural objects of their knowledge, especially as evinced in the distinction between mundane human experience and views of what nature "is really like"; third, the attempted mechanization of knowledge making, that is, the proposed deployment of explicitly formulated rules of method that aimed at disciplining the production of knowledge by managing or eliminating the effects of human passions and interests; and fourth, the aspiration to use the resulting reformed natural knowledge to achieve moral, social, and political ends, the condition of which was agreement that the knowledge in question truly was benign, powerful, and above all *disinterested.* The first and second themes are introduced in chapter 1; the third is treated mainly in chapters 2 and 3; and the fourth is almost exclusively handled in chapter 3.

Chapter 1 surveys some standard topics treated in most accounts of the Scientific Revolution: the modern challenge to Aristotelian natural philosophy and especially to the distinction between the physics appropriate for understanding terrestrial and celestial bodies; the attack upon an earth-centered, earth-static model and its replacement by the Copernican sun-centered system; the mechanical metaphor for nature, its association with mathematical means of understanding nature, and the "mathematization of qualities" manifested in the pervasive contrast between "primary" and "secondary" qualities.

The second chapter begins to depart from traditional ways of talking about the Scientific Revolution. It shifts attention from the body of knowledge treated simply as a product toward developing a more active and pragmatic sensibility about what it was like to *make* some scientific knowledge—what one had to do to secure and persuasively communicate a bit of natural knowledge. How did new knowledge differ in shape and texture from the old, and how did new knowledge-making practices differ from the old? I mean here to give readers a sense that the knowledge, and changes, described in the first chapter had to be laboriously made and justified, and to an extent, that practitioners diverged about how to go about securing and warranting natural knowledge. I want to introduce a dynamic sensibility toward science in action and science in the making rather than construing science as static and disembodied "belief."

A similar sensibility informs the last chapter, which aims to describe the range of historically situated *purposes* natural knowledge was put to in the seventeenth century. Natural knowledge was not just a matter of *belief;* it was also a resource in a range of practical activities. What did its advocates reckon a reformed natural philosophy was good for? What did they think could be done with it that could not be done with traditional forms of knowledge? Why should it be valued and supported by the other institutions of society?

While acknowledging the selective nature of this account, I want to intersperse interpretative generalizations with a series of relatively detailed vignettes of particular scientific beliefs and practices. I do this because I want this book, however arbitrarily selective, to give readers some feel for what it was like to have a certain kind of knowledge, to do a bit of natural knowledge making, to publicize and recognize its value in early modern society. I do not think this task has yet been satisfactorily attempted in a treatment of this purpose and scope. I mean the vignettes to serve as windows into the past, through which readers are invited to peek. I want to give at least a sense of early modern science not only as it was believed, but also as it was made and put to use. There is perhaps no more hackneyed historical intention than the wish to "make history come alive," yet it is something very like that desire that animates this book.

One

WHAT WAS KNOWN?

The Scope of Knowledge and the Nature of Nature

Sometime between the end of 1610 and the middle of 1611 the Italian mathematician and natural philosopher Galileo Galilei (1564–1642) trained the newly invented telescope on the sun and observed dark spots, apparently on its surface. Galileo reported that the spots were irregularly shaped and varied from day to day in number and opacity (fig. 1). Moreover, they did not remain stationary but appeared to move regularly across the disk of the sun from west to east. He did not profess to know with any certainty what these spots were made of. They might be physical features of the solar surface; they might be something similar to earthly clouds; or they might be "vapors raised from the earth and attracted to the sun." But whereas other contemporary observers reckoned that the spots were small planets orbiting the sun at some considerable distance from it, Galileo was sure, based on calculations in mathematical optics, that they were "not at all distant from its surface, but are either contiguous to it or separated by an interval so small as to be quite imperceptible."

Not Galileo's observations of sunspots but his particular interpretation of those spots was widely taken as a serious challenge to the whole edifice of traditional natural philosophy as it had been handed

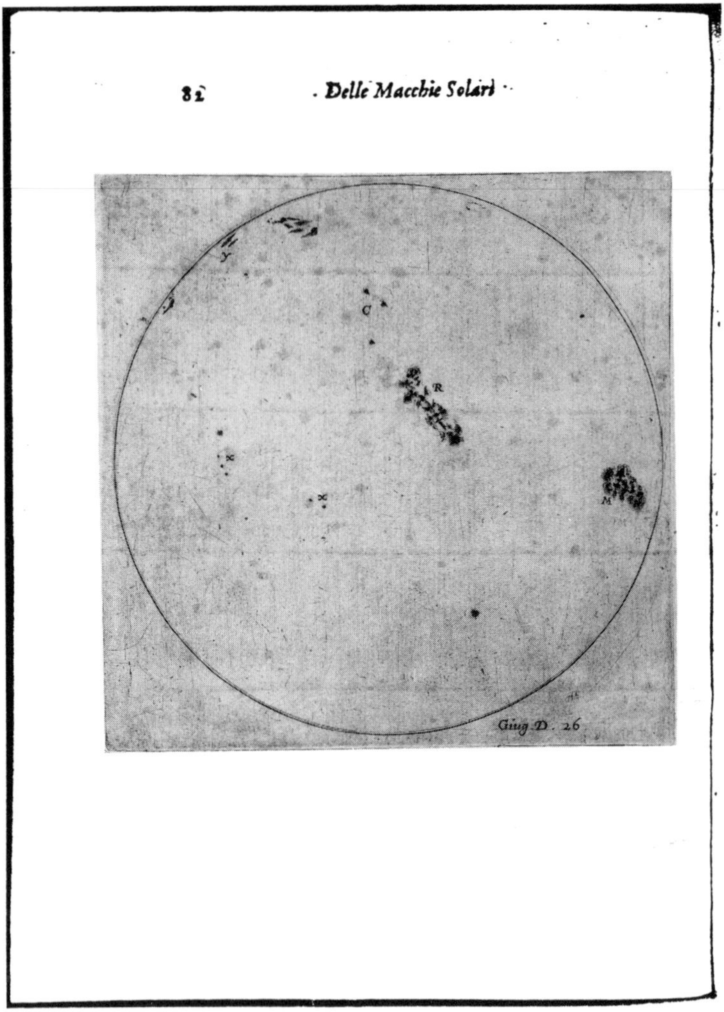

1. *Galileo's observations of sunspots on 26 June 1612. Source: Galileo Galilei,* Istoria e dimonstrazioni intorno alle macchie solari . . . *(Rome, 1613).*

down from Aristotle (384–322 B.C.) and modified by the Scholastic philosophers of the Middle Ages and Renaissance.[1] Galileo's views on sunspots, along with a body of other observations and theorizing, profoundly questioned a fundamental Aristotelian distinction between the physics of the heavens and that of the earth. Orthodox thinking, from antiquity to Galileo's time, had it that the physical nature and principles of heavenly bodies differed in character from those that obtained on earth. The earth, and the region between the earth and the moon, were subject to familiar processes of change and decay. All motion here was rectilinear and discontinuous. But the sun, the stars, and the planets obeyed quite different physical principles. In their domains there was no change and no imperfection. Heavenly bodies moved continuously and in circles, if they moved at all, uniform circular motion being the most perfect form possible. These are the reasons orthodox thinking located comets either in the earth's atmosphere or at least below the moon: these irregularly moving ephemeral bodies were just the sort of things that could not belong to the heavens. And though asserting the mutability of the heavens was not unknown in late sixteenth- and early seventeenth-century Aristotelian circles, making such a claim still strongly retained its status as a challenge to orthodoxy.

Within that orthodox framework the sun could not conceivably have spots or blemishes. Galileo was well aware of the sort of a priori reasoning that inferred from the traditionally accepted belief that the sun *was* immaculately and immutably perfect to the conclusion that the spots *could not* be on the solar surface. He argued against an Aristotelian opponent that it was simply illegitimate to take the sun's perfection as an undoubted premise in physical argument. Instead, we must move from what Galileo took as the observationally well supported fact that the spots were on the sun's surface to the conclusion

1. Scholasticism was a form of Aristotelian philosophy, especially as developed by Saint Thomas Aquinas (ca. 1225–74), and taught in the medieval universities ("Schools"). Adherents were sometimes called Schoolmen.

that there might be as much imperfection in the heavens as on the earth:

> It proves nothing to say . . . that it is unbelievable for dark spots to exist in the sun because the sun is a most lucid body. So long as men were in fact obliged to call the sun "most pure and most lucid," no shadows or impurities whatever had been perceived in it; but now it shows itself to us as partly impure and spotty, why should we not call it "spotted and not pure"? For names and attributes must be accommodated to the essence of things, and not the essence to the names, since things come first and names afterwards.

This was identified as a new way of thinking about the natural world and about how one ought to secure reliable knowledge of that world. Galileo was setting himself against traditionally accepted belief about the fundamental structure of nature, and he was arguing that orthodox doctrine ought not to be taken for granted in physical reasoning but should be made subject to the findings of reliable observation and mathematically disciplined reasoning.[2] So far as the possibilities of human knowledge were concerned, positions like Galileo's were profoundly optimistic. Like many others challenging ancient orthodoxy in the late sixteenth and early seventeenth centuries, Galileo was claiming that there existed not two sorts of natural knowledge, each appropriate to its proper physical domain, but only one universal knowledge. Moreover, by asserting the similarity of heavenly and terrestrial bodies, Galileo implied that studying the properties and motions of ordinary earthly bodies could afford understanding of what nature was like universally. It was not just that

2. The reliablity and authenticity of Galileo's telescopic observations—of the moon and the planets as well as of sunspots—were *not* in fact immediately conceded by all competent practitioners. There were substantial problems of persuasion involved in satisfying philosophers that, for example, the alleged phenomena were not illusions produced by the telescope, and chapter 2 will touch on some of these objections as well as problems attendant on the public authentication of observations made privately by an individual.

the imperfections and changeability of things on earth could be recruited as resources for understanding celestial phenomena; modern natural philosophers also claimed that earthly effects *artificially* produced by human beings could legitimately serve as tokens of how things were in nature. The motion of a cannonball could serve as a model for the motion of Venus.

Optimism about the possible scope of human knowledge was fueled by the new natural objects that were continually being brought to Europeans' attention. When Hamlet told Horatio that there were "more things in heaven and earth than are dreamt of in your philosophy," he was expressing sentiments similar to those of early modern natural philosophers challenging ancient orthodoxy. Traditional inventories of things that existed in the world were deemed to be illegitimately impoverished. What grounds were there for crediting ancient limits on the stock of factual knowledge? Every day new phenomena presented themselves about which the ancient texts were silent. Travelers from the New Worlds to east and west brought back plants, animals, and minerals that had no counterparts in European experience, and tales of still more. Sir Walter Raleigh protested to stay-at-home skeptics that "there are stranger things to be seen in the world than are contained between London and Staines."[3] From the early seventeenth century, observers using telescopes and microscopes claimed to reveal the limits of unassisted human senses and suggested that revelation of even more details and more marvels only awaited improved instruments. New and altered intellectual practices probed back in natural and human history and advanced claims to reliable knowledge about things no living person had witnessed. Newly observed entities that posed uncomfortable problems for existing philosophical systems were seized on by those eager to discomfit orthodox theorists. Who could confidently say

3. Staines was a village about twenty miles west of the City of London, near the present Heathrow Airport. Recent historical work has pointed out, however, that European experience of the New World was highly mediated through the long-standing textual traditions that generated expectations of what such a world *might be like*.

what did and did not exist in the world when tomorrow might reveal as yet undreamed-of inhabitants in the domains of the very distant and the very small?

In 1620 the English philosopher Sir Francis Bacon (1561–1626) published a text called *Instauratio magna* (*The Great Instauration*). The title itself promised a renovation of ancient authority, while the engraved title page was one of the most vivid iconographical statements of new optimism about the possibilities and the extent of scientific knowledge (fig. 2). A ship representing learning is shown sailing beyond the Pillars of Hercules—the Straits of Gibraltar that customarily symbolized the limits of human knowledge. Below the engraving is a prophetic quotation from the biblical Book of Daniel—"Many shall pass to and fro, and science shall be increased"—and Bacon later explained that the modern world had seen the fulfillment of the biblical prophecy when "the opening of the world by navigation and commerce and the further discovery of knowledge should meet in one time and place." The traditional expression of the limits on knowledge, *ne plus ultra*—"no farther"—was defiantly replaced with the modern *plus ultra*—"farther yet." The renovation of natural knowledge followed the enlargement of the natural world yet to be known. Practitioners of a mind to do so could use newly discovered entities and phenomena to radically unsettle existing philosophical schemes.

The Challenge to a Human-Centered Universe

Much of Galileo's astronomical and physical research in the early seventeenth century was undertaken to lend credibility to a new physical model of the cosmos that had first been published in 1543 by the Polish prelate Nicolaus Copernicus (1473–1543) (fig. 3). Until the middle of the sixteenth century no scholar in the Latin West had seriously and systematically questioned the system of Claudius Ptolemy (ca. A.D. 100–170) that placed an immobile earth at the center of the universe, with the planets, as well as the moon and the sun, orbiting

FRANCISCI
DE VERULAMIO,
Summi Angliæ
CANCELLARII,
Instauratio
magna.
Multi pertransibunt & augebitur scientia.
Sim: Pass: sculp:
LONDINI
Apud Joannem Billium
Typographum
Regium.
Anno
1620.

2. *The frontispiece of Francis Bacon's* The Great Instauration *(1620).*

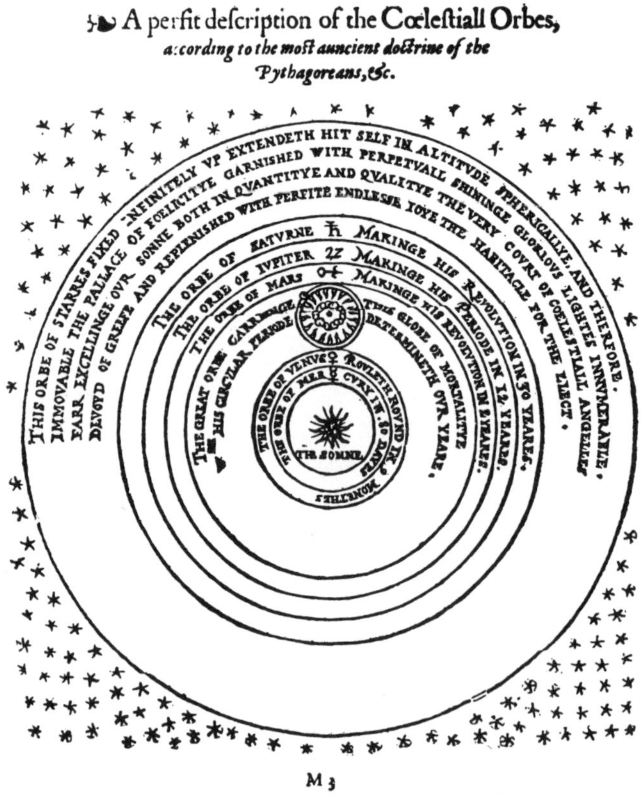

3. *The Copernican system, as depicted in the 1570s by the English mathematician Thomas Digges (ca. 1546–95). Digges modified Copernicus's views by developing a notion of a physical infinite universe in which the stars were placed at different points in that infinite space. Source: Thomas Digges,* A Perfit Description of the Caelestiall Orbes *(1576).*

in circles around the earth, each carried about on a physically real sphere (fig. 4). Farther out was the sphere that carried the fixed stars, and beyond that the sphere whose rotation caused the circular movement of the whole celestial system.

Ptolemy's geocentric system incorporated Greek views of the nature of matter. Each of the four "elements"—earth, water, air, and fire—had its "natural place," and when it was at that place it was at

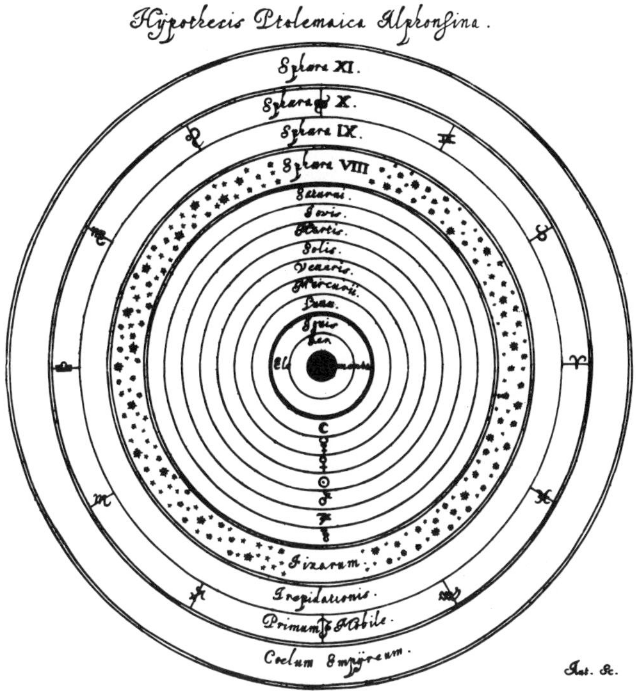

4. *The Ptolemaic cosmos, as depicted in the middle of the seventeenth century by the eminent German-Polish astronomer Johannes Hevelius (1611–97). Source: Johannes Hevelius,* Selenographia *(1647).*

rest. To be sure, all bodies we actually encounter on earth are not elementally pure, but what appears earthy has earth as a predominant element, the air we breathe has elemental air as its primary constituent, and so on. Earth and water are heavy elements, and they can be at rest only when they are at the center of the cosmos. Air and fire have a tendency to rise, and their proper spheres are above the earth. But heavenly bodies, including sun, stars, and planets, were made of a fifth element—the "quintessence" or "ether"—that was an incorruptible sort of matter, subject to different physical principles. So while earth tends to fall until it reaches the center of the universe, and air and fire tend to rise, the heavens and heavenly bodies naturally tend to move in perfect circles, and the stuff of which they are made is itself perfect and immutable.

The cosmos thus spun about the earth, the place where human beings lived, and in just that sense pre-Copernican cosmology was literally *anthropocentric.* Yet that quite special place did not necessarily connote special virtue. Although human beings, and their earthly environment, were understood to be the unique creations of the Judeo-Christian God, compared with the heavens and a heavenly afterlife the earth and earthly existence were regarded as miserable and corrupt, and the actual center of the cosmos was hell. In the late sixteenth century the French essayist and skeptic Michel de Montaigne (1533–92)—still accepting the Ptolemaic system—described the place where humans dwelled as "the filth and mire of the world, the worst, lowest, most lifeless part of the universe, the bottom story of the house." And even as late as 1640 an English supporter of Copernicanism recognized that a powerful current argument against heliocentrism proceeded from "the vileness of our earth, because it consists of a more sordid and base matter than any other part of the world; and therefore must be situated in the centre, and at the greatest distance from those purer incorruptible bodies, the heavens." Moreover, after Adam's and Eve's original sin and expulsion from Eden, human senses had been defiled, and the possibilities of human knowledge were understood to be severely limited. On the one hand, traditional thinking considered that the world in which humans spent their mortal lives—the world that was at the center of the universe—was uniquely changeable and imperfect; on the other hand, the scope and quality of the knowledge humans might attain were restricted.

The late sixteenth- and seventeenth-century natural philosophers who espoused and developed Copernicus's views attacked this anthropocentrism in fundamental ways. The earth was no longer at the center of the universe. Lifted into the heavens, it became merely one of the planets orbiting the sun, and in that quite literal physical sense, anthropocentrism was rejected.[4] The human experience of inhabiting a static platform, diurnally circled by sun and stars that were subject to their own annual motions, was denied. If common sense testified to the earth's stability, this new astronomy spoke of its double motion, daily about its axis and annually about the now static sun.[5] Common experience was here identified as but "appearance." If common sense expected that such motions, were they real, would cause people to hold onto their hats in the resulting wind or fall off the earth, then so much the worse for common sense. And if stones thrown straight upward tended to fall back to earth at the point they started from, then a new, noncommonsensical physics would be needed to show why this should happen on a moving earth. The earth's position in the universe was no longer unique. Some Copernicans even reckoned that this loss of uniqueness extended to the possibility that there were other inhabited globes and other types of humans, and in 1638 the English mathematician John Wilkins (1614–72) published a tract "to Prove that 'tis Probable there may be another habitable World" in the moon.

And if common human perception saw the earth canopied by a hemisphere of star-laden heavens, modern astronomers' accounts enormously extended the scale of the cosmos. When Galileo turned

4. There is another sense in which anthropocentrism was importantly *retained* within the new science of the seventeenth century. As chapter 3 will indicate, *mechanical* conceptions of nature conserved and supported a unique place for human beings within a created nature whose nonhuman parts were specially and divinely designed for human habitation and use. This kind of anthropocentrism remained central to science until the acceptance of Darwinism in the late nineteenth century.

5. In fact, Copernicus also posited a third motion for the earth: this was a very slow conical "wobble" of the earth's axis, and it was meant to account for small changes in observed stellar positions over thousands of years. A fully adequate account of astronomy in the Scientific Revolution would also treat the "compromise" between Ptolemy and Copernicus offered by the most skilled observational astronomer of the late sixteenth century, the Dane Tycho Brahe (1546–1601). The Tychonic system had the planets revolving about the sun and the sun revolving in turn about a stationary and central earth. In fact for many leading Copernicans the scheme to be opposed was Tycho's—favored by leading practitioners in the Catholic Jesuit order—rather than Ptolemy's.

his telescope to the stars he saw vastly larger numbers than were observable with the naked eye. To the three previously known stars in Orion's belt Galileo now added about eighty more (fig. 5). Some nebulous stars now were resolved into little Milky Ways. Galileo also noticed that, compared with the moon and the planets, stars did not appear to be much enlarged by the telescope. It was thus possible, though Galileo himself was reticent on the point, that the stars might be immensely far away. Such a view supported the Copernican system by accounting for the absence of parallax[6] that might otherwise be expected from a moving earth. Galileo's dramatic discovery of moons around Jupiter was used to give further credibility to the Copernican system, since the earth-moon relationship was no longer unique.

Traditional astronomy tended to posit a finite universe, each heavenly sphere revolving about the static earth and the whole of the heavens rotating once in twenty-four hours. In this system the stars could not be infinitely far away, for if they were, the sphere that carried them would have to move infinitely fast, and that was reckoned to be physically absurd. By contrast, Copernicus considered that the stars were fixed in space, and though he himself had insisted only that they were very far away, there was no longer any physical reason why the stars could not be infinitely removed. Some later advocates of the Copernican system did in fact stipulate that the sphere of the stars was "fixed infinitely up." So although the idea of an infinite universe had been broached in antiquity and though even several Copernicans bridled at it, the sixteenth and seventeenth centuries were the first periods in European culture when cosmic infinity seriously challenged the more comfortable dimensions of common experience. Human beings might occupy just a speck of dust in a universe of unimaginable size. And though many expert astronomers saw no rea-

6. Parallax is the change in angle when an object is viewed from two positions. The annual parallax of a close heavenly object ought to be noticeably large, whereas that for a very distant object might be so small as to be undetectable. Copernicus and his contemporaries could not detect any annual parallax for the fixed stars.

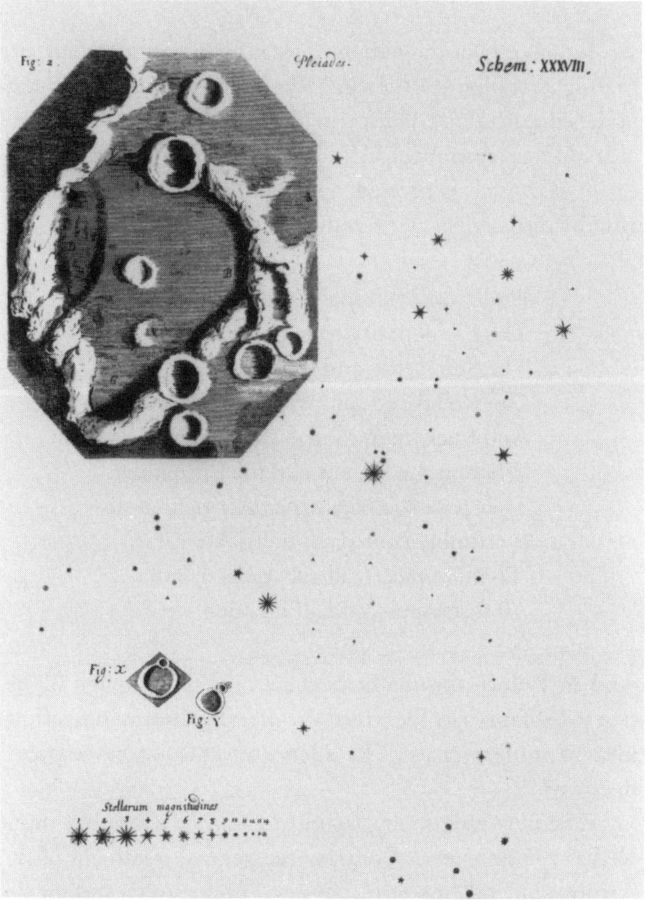

5. *"Of multitudes of small Stars discoverable by the Telescope." This illustration was included in the 1665* Micrographia *by the English experimentalist Robert Hooke (1635–1703). Only seven stars in the Pleiades are visible with the naked eye. Galileo's earlier telescope had been able to detect thirty-six. At the right and center, Hooke depicted seventy-eight stars he was then able to see with his twelve-foot telescope, their magnitudes indicated by the scale at the bottom left. This was taken as one indication of the rapidly increasing power of lens-assisted vision during the seventeenth century, and Hooke expressed confidence "that with longer Glasses . . . there might be discovered multitudes of other small Stars, yet inconspicuous."*

son for anxiety in the notion of an infinite cosmos (some even celebrating its sublimity), the same was not necessarily true for members of the educated laity. Unease in the face of infinity, of shaken systems of traditional cosmological knowledge, and of the decentering of the earth was widely expressed, nowhere more eloquently than it was in 1611 by the English cleric and poet John Donne:

> And New Philosophy calls all in doubt,
> The Element of fire is quite put out;
> The Sun is lost, and th' earth, and no man's wit
> Can well direct him where to look for it.
> And freely men confess that this world's spent,
> When in the Planets and the Firmament
> They seek so many new; then see that this
> Is crumbled out again to his Atomies.
> 'Tis all in pieces, all coherence gone;
> All just supply, and all Relation.

And in France the mathematician and philosopher Blaise Pascal (1623–62) famously identified the morally disorienting effects of the idea of infinite space: "Le silence éternel de ces espaces infinis m'effraye."[7]

The new philosophy assaulted common sense at a mundane as well as a cosmic level. Consider the general treatment of motion in Aristotelian and "modern" physics. For Aristotle, and for those medieval and early modern philosophers who followed him, the elements of earth, water, air, and fire each had its "natural motion," the way it was "in its nature" to move. As we have seen, for the element of earth the natural motion was to descend in a straight line toward the center of the earth, and this it will do unless the earthy body encounters either an obstacle that blocks its path or a push that acts on it

7. "The eternal silence of infinite space frightens me." These words were meant to express not Pascal's own attitudes as a philosopher but those of contemporary "libertines."

in another direction. Natural motion tends toward natural place. Aristotle was, of course, well aware that all sorts of nonrectilinear motions occurred. These were called "violent motions," motions against the nature of a body, to be accounted for by the action of external forces, such as might be imposed on a stone by a person's throwing it upward or parallel to the ground. But we cannot learn about natural motions by considering those motions artifically forced on a body.

So for Aristotle and his followers all natural motion had a developmental character. Bodies naturally moved so as to fulfill their natures, to transform the potential into the actual, to move toward where it was natural for them to be. Aristotelian physics was in that sense modeled on biology and employed explanatory categories similar to those used to comprehend living things. Just as the acorn's development into the oak was the transformation of what was potential into what was actual, so the fall of an elevated stone was the actualization of its potential, the realization of its "nature." The resonance between traditional accounts of natural motion and the texture of human experience is evident. Human beings offered teleological—or goal-orientated—accounts of their own movements. Why does the shepherd move toward his cottage? Because he forms a purpose to be where he wishes. Why do the flames leap up out of the fire? Because they aspire to be at their natural place. It is in just this sense that traditional physics on the eve of the Scientific Revolution had a human-scaled character. The basic character of the categories used to explain how rocks move was recognizably similar to that of those used to account for how we move. For that reason one may loosely refer to such traditional views of matter as "animistic," attributing soul-like properties (the Latin *anima* means soul) to natural objects and processes.[8]

8. Historians have alternatively referred to such patterns of belief as *hylozoist,* a compound deriving from the Greek words for matter and life. The reference to the human-scaled nature of Aristotelian physics partly reflects a characterization polemically developed by its seventeenth-century opponents. Although the point about resonances between human and natural explanatory categories stands, it is important to note that Aristotle himself warned against the idea that "nature deliberates."

It was these teleological and animistic features of the traditional physics of motion that the new natural philosophers of the seventeenth century seized on—indeed, caricatured—as marks of its absurdity and *unintelligibility.* What had given physics its grip on common sense for centuries was now to be seen as a sign of its inadequacy. Just to state the teleological character of Aristotelian natural philosophy was to count as critique. The English philosopher Thomas Hobbes (1588–1679) was one of many seventeenth-century critics of Aristotelianism who discredited traditional physical beliefs by drawing sarcastic attention to their anthropomorphism. Aristotelians said that bodies descended because they were heavy: "But if you ask what they mean by *heaviness,* they will define it to be an endeavour to go to the centre of the earth. So that the cause why things sink downward, is an endeavour to be below: which is as much to say that bodies descend, or ascend, because they do. . . . [It is] as if stones and metals had a desire, or could discern the place they would be at, as man does."

The Natural Machine

The framework that modern natural philosophers preferred to Aristotelian teleology was one that explicitly modeled nature on the characteristics of a *machine.* So central was the machine metaphor to important strands of new science that many exponents liked to refer to their practice as the *mechanical philosophy.* Modern practitioners disputed the nature and the limits of mechanical explanation, but *proper* mechanical accounts of nature were widely recognized as the goal and the prize. Yet the very idea of construing nature as a machine, and using understandings derived from machines to interpret the physical structure of nature, counted as a violation of one of the most basic distinctions of Aristotelian philosophy. This was the contrast between what was natural and what was contrived or artificial.

The conception of nature as an artificer was far from unknown in Greek and Roman thought and was, indeed, prominent in Aristotle's *Physics*. Nature carries out a plan, just as a human architect constructing a house, or an armorer making a shield, intentionally executes a plan. Because both natural and human work may be regarded as artifice, there *are* grounds for specific comparison: so one may say, with the Greeks, that art (here meaning artifice or technology) imitates nature. Human art may assist, complete, or modify nature—as in agriculture—or it may frankly imitate nature—as does the human spinner or weaver emulating the spider's work. (Other ancient philosophers said that the art of cooking imitated the sun and that machine making was inspired by observation of the rotating heavens.) However, it was not proper to suppose that the artifice of nature and that of humans belonged on the same plane. Nature, though capable of making mistakes, was far superior to human artifice, and it was impossible that humans should compete with nature. Any such ambition might also be considered immoral, for the world order is divine and humans' pretensions to do what divinity did were illicit. Roman writers told stories of the Golden Age, when humans lived happily and satisfactorily without architects, weavers, or even, in some versions, agriculture. As natural and human artifice were compared, so they were *opposed*. And the grounds of their opposition in traditional thought told against the legitimacy of using artificial devices either to interrogate or to model the natural order.

Nevertheless, the precondition for the intelligibility and the practical possibility of a mechanical philosophy of nature was setting aside that Aristotelian distinction, as it had been developed and protected through the Middle Ages and the Renaissance. Such writers as Bacon made that rejection the basis for both a reformed natural history—now to include the products of human artifice—and a more optimistic attitude toward the potential of human artifice: "The artificial does not differ from the natural in form or essence . . . nor matters it, provided things are put in the way to produce an effect, whether it be done by human means or otherwise." This Baconian sensibility was widely endorsed by seventeenth-century mechanical philosophers. In France the atomist Pierre Gassendi (1592–1655) wrote that "concerning natural things, we investigate in the same way as we investigate things of which we ourselves are the authors." And the French mathematician and philosopher René Descartes (1596–1650) announced that "there is no difference between the machines built by artisans and the diverse bodies that nature alone composes" except that the former must necessarily be proportioned in size to the hands of their builders, whereas the machines that produce natural effects may be so small as to be invisible. "It is certain," Descartes wrote, "that there are no rules in mechanics which do not hold good in physics, of which mechanics forms a part or special case (so that all that is artificial is also natural); for it is not less natural for a clock, made of the requisite number of wheels, to indicate the hours, than for a tree which has sprung from this or that seed, to produce a particular fruit." The heat of the sun can be legitimately compared to terrestrial fire; the gold said to be produced by the alchemist is the same as that found naturally in the earth; the physics appropriate to understanding machines made by humans may be the same as that required for understanding celestial motions; and as we shall see, the causes of *all* sensible natural effects may be treated as flowing from the actions of "micromachines." It was a widespread seventeenth-century sentiment that humans can securely know *only* what they themselves construct by hand or model by mind.

Of all the mechanical constructions whose characteristics might serve as a model for the natural world, it was the *clock* more than any other that appealed to many early modern natural philosophers. Indeed, to follow the *clock metaphor* for nature through the culture of early modern Europe is to trace the main contours of the mechanical philosophy, and therefore of much of what has been traditionally construed as central to the Scientific Revolution. Mechanical clocks were present in Europe by the late thirteenth century, and by the middle of the fourteenth century weight-driven mechanical clocks had become a fairly standard feature of larger cities. Early clocks typically had their workings exposed to full view, and consequently the relation between the movements of hands indicating time and the mechanical means by which these movements were produced was well understood. By the sixteenth century, however, the tendency developed to house the clockworks in opaque boxes, so that only the time-telling movements, not their mechanical means of production, were routinely visible. Public clocks became more and more complex in the effects they could produce and more and more integrated into the practical life of the community. So, for example, whereas traditional temporal "hours" measured by the sundial might vary in length according to the season and the latitude, the hours told by the mechanical clock were constant over space and time, taking no heed of the natural rhythms of the universe or of the varying situated practices of human life. The patterns of human activity might now be regulated according to mechanical time rather than time being told in relation to the rhythms of human life or natural movements.

For those sectors of European society for whom the clock and its regulatory functions were important aspects of daily experience, this machine came to offer a metaphor of enormous power, comprehensibility, and consequence. The allure of the machine, and especially the mechanical clock, as a uniquely intelligible and proper metaphor for explaining natural processes not only broadly follows the contours of daily experience with such devices but also recognizes their potency and legitimacy in ordering human affairs. That is to say, if we want ultimately to understand the appeal of mechanical metaphors in the new scientific practices—and the consequent rejection of the distinction between nature and art—we shall ultimately have to understand the power relations of an early modern European society whose patterns of living, producing, and political ordering were undergoing massive changes as feudalism gave way to early capitalism.

In 1605 the German astronomer Johannes Kepler (1571–1630) announced his conversion from his former belief that "the motor cause" of planetary motion "was a soul": "I am much occupied with the investigation of the physical causes. My aim in this is to show that the machine of the universe is not similar to a divine animated being, but similar to a clock." In the 1630s Descartes elaborated a set of extended causal analogies between the movements of mechanical clocks and those of all natural bodies, not excepting even the movements of the human body: "We see that clocks . . . and other machines of this kind, although they have been built by men, do not for this reason lack the power to move by themselves in diverse ways." Why shouldn't human respiration, digestion, locomotion, and sensation be accounted for in just the way we explain the motions of a clock, an artificial fountain, or a mill? In the 1660s the English mechanical philosopher Robert Boyle (1627–91) wrote that the natural world was "as it were, a great piece of clock-work." Just as the spectacular late sixteenth-century clock in the cathedral at Strasbourg (fig. 6) used mechanical parts and movements to mimic the complex motions of the (geocentric) cosmos, so Boyle, Descartes, and other mechanical philosophers recommended the clock metaphor as a philosophically legitimate way of understanding how the natural world was put together and how it functioned. For Boyle the analogy between the universe and the Strasbourg clock was both exact and fertile: "The several pieces making up that curious engine are so framed and adapted, and are put into such a motion, that though the numerous wheels, and other parts of it, move several ways, and that without any thing either of knowledge or design; yet each part performs its part in order to the various ends, for which it was contrived, as regularly and uniformly as if it knew and were concerned to do its duty."

A number of features of the clock thus struck many seventeenth-century mechanical philosophers as appropriate metaphorical resources for understanding nature. First, the mechanical clock was a complex artifact designed and constructed by people to fulfill functions intended by people. Although it was itself inanimate, the clock imitated the complexity and the purposiveness of intelligent agents. If you did not know there was an intelligent clockmaker who purposefully brought it into being, you might suppose that the clock itself was intelligent and purposive. The contemporary popularity of automatons—machines that vividly mimicked the motions of animals and humans—also impressed a number of mechanical philosophers (see the cock automaton in fig. 6). That skillfully contrived machines might trick naive observers into believing they were seeing

6. *The Strasbourg cathedral clock. The second Strasbourg clock Boyle referred to was completed in 1574. This illustration shows the clock as reconstructed in the 1870s. It not only tells time but also indicates solar and lunar cycles, calculates eclipses, and so on. The cock automaton on the top of the tower at the left crows thrice every day at noon in memory of the temptation of Saint Peter. Source:* Scientific American, *10 April 1875.*

something natural and animate counted toward the legitimacy of the mechanical metaphor. Yet one thing that competent people reliably know about clocks and automatons is that they are *not* intelligent agents. Hence the clock, and similar mechanical contrivances, provided valuable resources for those concerned to provide a convincing alternative to philosophical systems that built intelligence and purpose into their schemes of how nature worked. Machines might be *like* purposive agents and might even substitute for purposive human labor, and that likeness constituted part of their metaphorical appeal. Yet they were competently understood not to be purposive agents, and that difference constituted part of their explanatory power. You could get the appearance of complex design and purpose in nature without attributing design and purpose to material nature. There might be an intelligent agent in the universe standing in the same relation to nature as clockmakers did to their clocks, as we shall see in chapter 3, but one was not to confuse the inanimate product of intelligence with intelligence itself.

The clock was also an exemplar of uniformity and regularity. If philosophers saw the natural world as exhibiting orderly patterns of movement, then the mechanical clock was available as a model of how regular natural motions might be mechanically produced. Machines in general had a determinate structure: the materials and motions required to make them, and to make them go, were knowable by human beings and, in principle, specifiable. That is to say, machines were accounted wholly *intelligible.* In that culture it was represented that there was nothing mysterious or magical, nothing unpredictable, nothing causally capricious about a machine. The machine metaphor might, then, be a vehicle for "taking the wonder out" of our understanding of nature or, as the sociologist Max Weber put it in the early twentieth century, for "the disenchantment of the world." Machines thus provided a model of the form and scope that human knowledge of nature might properly have and of how human accounts of nature might properly be framed. Think of nature as if it were a machine; attend to the uniformities of its motions and not to the occasional irregularities that can be observed even in the best-made machines; give interpretations of nature, so far as possible, as if it were a causally specifiable engine. Accounts of nature that take this form can be thought of as philosophically proper, legitimate, and intelligible.

It must, however, be pointed out that there is nothing, so to speak, "in the nature" of machines to prevent them from being regarded as mysterious, and a strand of thought going back to the Hellenistic period accounted machines something more than the sum of their material parts. Boyle, for example, wrote about the cultural variability of appreciations of machinery. He related a—probably apocryphal—story about the Jesuits "that are said to have presented the first watch to the king of *China,* who took it to be a living creation." Boyle himself accepted the adequacy of an account wholly in terms of "the shape, size, motion &c. of the spring-wheels, balance, and other parts of the watch," while recognizing that he "could not have brought an argument to convince the Chinese monarch, that it was not endowed with life." A mechanical metaphor for nature meant, as all metaphors accepted as legitimate do, that our understanding of *both* terms changes through their juxtaposition. The rightness of a metaphor is not subject to proof.

For philosophers of Boyle's and Descartes's disposition a mechanical account of nature was thus explicitly contrasted with the anthropomorphism and animism of much traditional natural philosophy. To do mechanical philosophy was therefore to be seen to be doing something radically different from attributing purpose, intention, or sentience to natural entities. Mechanical accounts of natural phenomena varied widely. Some philosophers ventured to say more than others about the mechanical constitution of nature, and later sections of this book will discuss *what it meant* to give a mechanical explanation of natural phenomena, what the limits of such explanations were supposed to be, and what domains were considered appropriate for mechanical accounts. Yet, despite this variation, all seventeenth-century mechanical accounts set themselves in opposition to the tradition that ascribed to nature and its components the capacities of purpose, intention, or sentience.

It was well known in the seventeenth century that suction pumps could not be made to raise water to a height of more than about thirty-three feet (fig. 7). This inability was attributed partly to problems with the materials used—for example, the porosity of wooden pipes—and partly to the traditional doctrine that nature abhors a vacuum.[9] That a suction pump could draw water up at all was traditionally taken to depend on water's abhorrence of a vacuum, its attempt to rise up to prevent a vacuum from forming at the top, while the limited height of the column might be treated as a quantitative measure of the strength of that abhorrence. Consequently the traditional explanation of a well-known, and practically important, effect was explained by ascribing purposelike characteristics to a bit of nature, in this case to a quantity of water.

The problems posed by the phenomena of suction pumps formed a centerpiece of the distinction between "new" and "old," "mechanical" and "Aristotelian" philosophies of nature. In 1644 an admirer of Galileo, the Italian mathematician Evangelista Torricelli (1608–47), attempted to explain pumplike effects better and, specifically, to test the validity of a mechanical account that did not attribute to fluids anything like a capacity to abhor. Suppose that the height of liquids in suction pumps had to do not with the existence or strength of "abhorrence" but with a simple mechanical equivalence in nature. Inside the pump one had a column of water, outside a column of atmospheric air. The column of water reached its resting height when its weight equaled the weight of the atmospheric air pushing against its base. Torricelli thus sought to model the phenomena of pumps on the well-understood workings of a mechanical balance. In fact, the view that air had a weight was in itself a challenge to traditional "natural place" beliefs, since Aristotelians maintained that neither air nor

9. Many, though not all, ancient natural philosophers regarded the idea of a vacuum in nature as an impossibility. Certainly this was the influential view of Aristotle, and seventeenth-century mechanical philosophers were divided on whether vacuums were possible or whether nature was full of matter, a plenum.

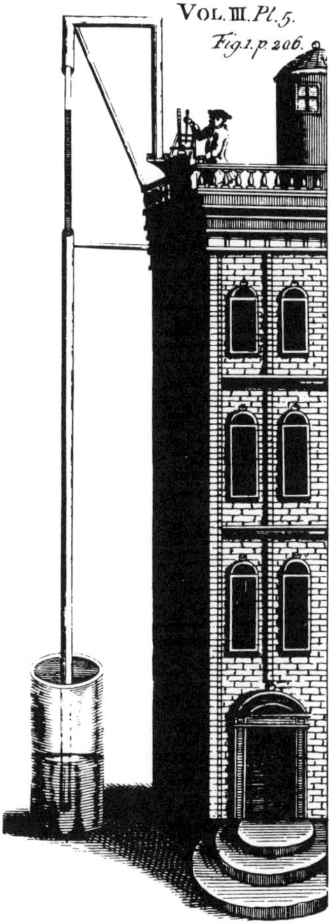

7. The raising of water by Robert Boyle's suction pump. When this experiment was made in the 1660s, the limit of about thirty-three feet to which pumps could raise water was already widely known in both artisanal and philosophical circles. This had been established in the 1640s by Gasparo Berti, who was in turn inspired by remarks in Galileo's Two New Sciences *of 1638. Boyle wanted to assure himself of the matter of fact, suspecting that previous apparatus was "not sufficiently staunch, nor the operation critically enough performed and taken notice of." For ease of observation, the upper two or three feet of the tube was made of glass, fastened by cement to the metal lower section of about thirty-two feet. Boyle's pump raised the water to a maximum height of thirty-three and a half feet. The house used for this purpose was probably close to Boyle's own Pall Mall residence in London. Source: Robert Boyle,* Continuation of New Experiments Physico-mechanical Touching the Spring and Weight of the Air *(1669).*

water weighed upon itself "in its proper place," for example, air in the atmosphere and water in the sea.

Mercury was known to be about fourteen times as dense as water. Accordingly, a mechanical account predicted that if a glass tube, sealed at one end, was filled with mercury and then inverted in a basin of mercury, the resting level of the mercury ought to be only

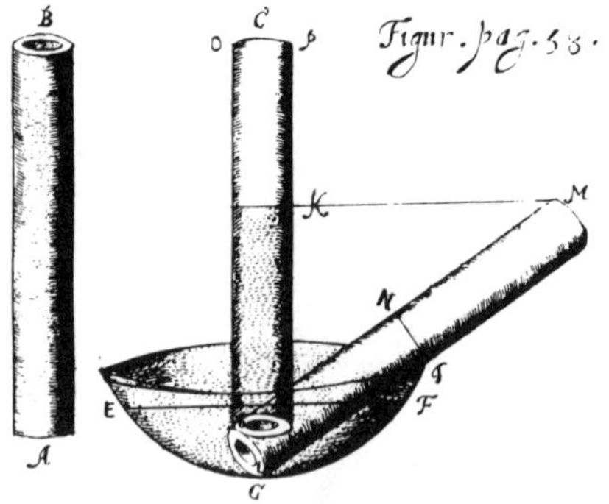

A B, *A Tube of Glass, replete with Quickſilver.*
A, *The lower extreme thereof, hermetically ſealed.*
B, *The upper extreme thereof; open.*
D C, *The ſame Tube inverted, and perpendicularly erected in a veſſel full of Quickſilver: ſo as the orifice* D, *be not unſtopped, untill it be immerſed in the ſubjacent Quickſilver.*
H G I, *A veſſel filled up to the line* E F, *with Quickſilver: and thence up to the brim* H I, *with Water.*
C K, *The Vacuum, or Space deſerted by the Quickſilver deſcended.*
O C P, *The quantity of Aer ſuppoſed to have inſinuated it ſelf at the ſubduction of the finger from the inferior orifice* D.
K M, *A Line parallel to the Horizon.*
L M, *The ſame Tube again filled with Quickſilver, and reclined untill the upper extreme thereof become parallel to the ſame horizontal altitude with* K.
N, *The diſtance of* 27 *inches from* L, *as* K *from* D.

8. *This image of the Torricellian experiment comes from Walter Charleton's* Physiologia Epicuro-Gassendo-Charltoniana *(1654), a seminal work in the seventeenth-century revival of Greek and Latin atomism. It illustrates a version of the phenomenon originally displayed by Evangelista Torricelli in the 1640s. Charleton (1620–1707) was arguing here that the space above the mercury column is free of air. When the tube is inclined from the perpendicular (right), the level of mercury fills up the previously existing space, and Charleton rhetorically asked where any air supposed to exist in that space could go, since the tube was hermetically sealed at the top and no bubbles could be seen to pass through the mercury.*

one-fourteenth as high as the level reached by water in suction pumps. And this was what was observed (fig. 8). "We live," Torricelli announced, "at the bottom of an ocean of the element air, which by unquestioned experience is known to have weight." Torricelli had in fact constructed the first *barometer*—from the Greek words for weight and measure—in what many regarded as a decisive confirmation of the mechanical view of nature. Many but by no means all. The view that an abhorrence of a vacuum played *some* legitimate role in explaining such results was well entrenched and thought to be plausible by very many early to mid-seventeenth-century philosophers who were otherwise well disposed toward mechanism: Galileo himself was of this opinion.

In France, Pascal initially reckoned that the Torricellian experiment proved only that the force of nature's abhorrence of a vacuum was finite. All that Torricelli had established was that its force was measured equally well by thirty-three feet of water and twenty-nine inches of mercury. Lacking confidence in generalizing about nature from some few artificially produced effects, Pascal was not inclined to accept the analogy with a mechanical balance unless he could vary the weights *on both sides*. Late in 1647, Pascal asked his brother-in-law Florin Périer to carry the Torricellian barometer up the volcanic peak in central France called the Puy de Dôme and to observe what changes, if any, greater elevation produced in the level of mercury. When the ascent was eventually made in September 1648, a similar barometer was left in the care of a monk at a convent at the base of the mountain so that the recorded mercury level in it could act as a "control." The brother-in-law reported that the mercury level at the top of the peak—approximately three thousand feet above the starting point—was about three inches lower. Less of the atmosphere was weighing down on the barometer at the summit than on the one at the foot of the mountain. The barometer's behavior was taken to be caused by the weight of the air and in turn to be a reliable measure of that weight. Pascal, accordingly, announced his conversion to the mechanical view: "All the effects ascribed to [the abhorrence of a vacuum] are due to the weight and pressure of the air, which is their only real cause."[10] To be a mechanical philosopher was to prefer inani-

10. The Puy de Dôme experiment was repeated several times by other practitioners climbing other mountains. Although the original experiment was evidently decisive for Pascal, others could not replicate the fall in the mercury level. Nor were

mate interpretations like the weight of air to the implied intentionality of matter's abhorrence.

Many mechanical philosophers favorably contrasted their accounts of natural phenomena to those that invoked "occult" powers. In the Renaissance "natural magical" tradition, for example, it was common to suppose that bodies might act on each other at a distance through occult powers of sympathy, attraction, or repulsion. Although the effects of such powers were regarded as observable, the means by which they acted were hidden (which is why they were called occult) and might not be specifiable in terms of the ordinary "manifest" properties of sensible matter. Thus it was by invoking occult powers that astrological influences from celestial bodies like the planets were said to act on on earthly affairs, that the sun had the capacity to bleach, that rhubarb could act as a laxative, and that the magnet attracted iron. These powers were all said to be perceptible from their effects but could not be inferred from the manifest appearance of planets, sun, rhubarb, or magnets.[11] The human body (the microcosm) was connected to the universe (the macrocosm) through a series of occult correspondences and influences. By no means all new philosophers sought to discredit the legitimacy of occult powers, nor did all of them reject at least some of the claims of the astrological tradition. Among astronomers, Kepler and his contemporary Tycho Brahe were astrological adepts; and Bacon and Boyle, for example,

---

resources absent to account for an observed fall without accepting complete mechanism, for example, by pointing to the possible role of temperature changes. A distinction between the weight and the pressure of the air will be treated in the next chapter.

11. The meanings of the word "occult" varied and changed in the early modern period. Moreover, the description of explanatory accounts *as* occult was widely used by mechanical philosophers as a form of accusation. For example, mechanically inclined practitioners who refused to offer a specific causal account of how a certain physical effect was produced might be accused by others of reintroducing discredited occult powers, as was the case in the early eighteenth-century disputes over gravitation between Newton and Leibniz noted later in this chapter. It has even been argued recently that, by shifting the meaning of occult qualities from what was hidden and insensible to what was visible in its effects but *unintelligible* in mechanical and corpuscular terms, modern natural philosophers actually reintroduced occult qualities while claiming to reject them.

accepted the principle of natural celestial influences wholeheartedly while expressing skepticism about some more ambitiously specific predictive forms of astrology. Boyle and other fellows of the Royal Society of London in the 1660s and 1670s had no doubt that disembodied spirits, witches, and demons exerted effects in the natural world, although, as we shall see, their place in mechanical philosophy and the means used to establish the veracity of particular claims were subject to contestation and control. But it was nevertheless characteristic of the new practice to express suspicion about a range of empirical claims of occult influence and, in others, to seek to translate them into material and mechanical terms.

Although the mechanical philosophy developed strongly in opposition to Aristotelian doctrine, the tradition of "Renaissance naturalism" also provided an important model of what was to be opposed. This "naturalism" was taken to be deeply rooted in the overall culture, and many of those attracted to mechanism were disturbed by what were seen as the consequences of naturalism for a wide range of valued cultural and social practices through the seventeenth century and into the eighteenth. It was partly through these processes of opposition that the mechanical conception of nature emerged and was sustained. In early seventeenth-century France, for example, the philosopher and mathematician Father Marin Mersenne (1588–1648) of the Catholic order of Minims saw very dangerous consequences flowing from the Renaissance revival of the doctrine of the anima mundi, or world soul—the notion that matter was imbued with life and the associated identification of God with nature. Such doctrines gave legitimacy to magical beliefs and practices, and also, as Mersenne especially feared, to religious heresy. Mersenne worried that projecting supernatural powers onto things that, properly speaking, do not have such powers would blur the religiously crucial distinction between the natural and the supernatural—to the ultimate detriment of Christian belief and Christian institutions.

By imbuing the natural world with a range of inherent active powers, Renaissance naturalism tended to dispense with the explanatory role of God, rightly conceived as the one wholly supernatural entity. That was what, more than anything else, had to be opposed in the name of proper Christian religion. Although Aristotelianism possessed resources that were in principle valuable for combating naturalism—for example, its endorsement of the immortality of the soul and its rejection of determinism—its record of effectiveness in responding to the challenge of Renaissance naturalism had not been good. Nor had Aristotelianism offered convincing explanations of the sorts of phenomena much traded in by the naturalists, for example, magnetic attraction and herbal healing. In Mersenne's thinking the root problem was the idea of matter as essentially active, and the root solution was to be an account of matter as completely passive and inert—in other words, a *metaphysics* appropriate to a mechanical account of the natural world.[12] Given such an account of passive matter, the consequential distinctions between what was natural and what was supernatural could be maintained. Indeed, the presumption of passive matter was fundamental to a range of seventeenth-century versions of mechanism. Mersenne's influence in the development of mechanism and its appropriate view of matter was considerable. It was elaborated by his friend Descartes in the 1630s and 1640s, and through Descartes it was taken up, with modifications, by Hobbes, Boyle, and many others. And though the view of matter-as-passive was central to mechanical natural philosophy throughout the seventeenth century, it remained intermittently under attack from both philosophical and lay sources, and the ways that view worked out in specific explanatory tasks were highly varied.

It was part of the mechanists' credo that all genuine effects in nature were to prove explicable based on ordinary, comprehensible mechanical and material causes. So Bacon was suspicious of the

12. *Metaphysics* is the philosophical inquiry into "first principles," including the attempt to characterize the ultimate nature of what exists in the world. Although some modern writers regarded metaphysics as an important part of natural philosophy, or even as its foundation, others condemned metaphysical speculation as beyond the proper bounds of scientific inquiry, using the term metaphysics as a loose pejorative for philosophical claims that were abstruse, abstract, or otherwise undecidable by ordinary means.

claimed matter of fact involved in the celebrated "weapon salve." It had been commonly avowed that a wound caused by a given weapon might be healed by anointing not the wound itself but the sword or knife that caused it, even when wound and weapon were up to thirty miles apart. Bacon was not yet prepared either to accept or to reject the reality of the claimed effect, and he suggested that it be subjected to more critical trial. It *might be* amenable to explanation in material terms, but first the real existence of the effect had to be established.[13] Bacon also surveyed "many ancient and received traditions and observations touching the sympathy and antipathy of plants." It was common to explain why some plants thrived best when growing near other types by invoking occult principles of sympathy. Here he flatly repudiated "any such secret friendship and hatred" as "utterly mistaken" and suggested that some genuine effects might be referred to mundane causes—the effects on plants of neighbors' drawing certain nutrients out of the soil.

In the mid-1660s English physicians and natural philosophers disputed both the reality and the proper explanation of alleged "stroking" cures attributed to the Irish healer Valentine Greatrakes. Many trustworthy sources testified that Greatrakes had cured sufferers from scrofula, ulcers, and kidney stones by the laying on of his hands. Boyle's approach to such "stupendous performances" was to cautiously credit their reality and offer a provisional mechanical interpretation of how they might actually work. He said that he was not convinced there was anything "purely supernatural" about the cures and endeavored "to give a physical account" of them. Perhaps material "sanative effluvia" passed from Greatrakes's body to that of the patient, and perhaps it was these effluvia that effected the cure.

13. If Bacon was skeptical of the matter of fact, the weapon salve (or "powder of sympathy") had influential advocates in the seventeenth century. The English courtier, philosopher, and eventually fellow of the Royal Society of London, Sir Kenelm Digby (1603–65) was satisfied of its efficacy—using it himself to cure a wounded duelist whom the king's surgeons could not help. And Digby also offered an account of *how* it worked that mixed the resources of mechanism and occult sympathy in a manner not uncommon in seventeenth-century philosophy.

The emission of material effluvia and the effects performed thereby were said to be wholly explicable on mechanical principles. Nothing occult or supernatural need be adduced. If there was to be wonder attached to stroking cures, it ought to be directed toward the mechanical causes acting in God's nature, not toward things identified as mysterious and immaterial.

The Mathematization of Qualities

In Boyle's summary there were only "two grand principles" of the mechanical philosophy: matter and motion. There were no principles more primary, more simple, more comprehensive, and more comprehensible. Matter and motion were like the letters of the alphabet, simple and finite in themselves but capable in combination of producing almost endless diversity. So far as it concerned a properly conceived practice of natural philosophy, everything in the natural world was to be explained with reference to the irreducible properties of matter and its states of motion: that was one thing that made the interpretation of nature like the interpretation of machines. Nothing occult was supposed to be involved in talking about matter and motion. A mechanical account of nature was then given its limiting form and content: specify the shape, size, arrangement, and motion of the material constituents of the things concerned.

Seventeenth-century mechanical philosophers traced the legitimacy of such a view of nature's fundamental structure back to scriptural sources. The apocryphal Wisdom of Solomon handed it down that God "has disposed of all things in number, weight, and measure," and similar sentiments were intermittently expressed throughout the Middle Ages. What was new in the seventeenth century was the vigor with which the principles of matter and motion were advanced as defining resources of a proper natural philosophy. If a purportedly natural philosophical account brought to bear resources other than matter and motion, it ran a substantial risk of being identified as unintelligible, as not in fact being philosophical at all.

Despite this basic agreement among mechanical philosophers, the specificity and content of mechanical accounts of particular natural phenomena varied considerably from one practitioner to another. Descartes preferred to spell out in detail how the size, shape, motion, and forms of interaction of insensible bits of matter might produce all the diversity of physical effects. He supposed that all physical bodies were composed of three "elements," comprising the same basic kind of matter but differing in size and shape. In order of particle size, the elements ascended from what he sometimes called "fire" (the smallest) to "air" to "earth" (the largest).[14] Some bodies—for example, the sun and fixed stars—were "pure," consisting of just the one element of fire; others were of "mixed" composition, for example, all the objects we encounter in terrestrial environments, including animate bodies. Cartesian physical explanations, then, consisted of specifying bodies' particulate composition and the particles' states of motion.

Magnetism, for instance, was explained by reference to screwlike particles generated by a vortex around the earth, which particles fit into appropriately configured pores in iron (fig. 9). The motion of these particles streaming between a magnet and a piece of iron forced away the air between the two bodies and thus drew them together. The existence of dual magnetic poles was accounted for by positing left- and right-handed screws. Similarly, the human body might be treated as if it were but "an earthen machine." Digestion was the heat-induced separation of food particles, the coarsest descending, ultimately to be expelled through the rectum, and the finest particles

14. Descartes, like the Aristotelians, did not admit the existence of a vacuum in nature, and accordingly he stipulated that particles of the first element did not have a determinate size and shape but were able to divide and change shape on collision "in order to accommodate themselves to the spaces they enter." His insistence on the indefinite divisibility of the particles of matter thus distinguished his matter theory from that of such contemporary "atomists" as Pierre Gassendi and Gassendi's important English exponent Walter Charleton. To hold to a corpuscular or particulate view of matter was therefore not necessarily the same thing as to maintain the defining doctrine of atomism: that all bodies were made up of invisible, impenetrable, and *indivisible* bits of matter.

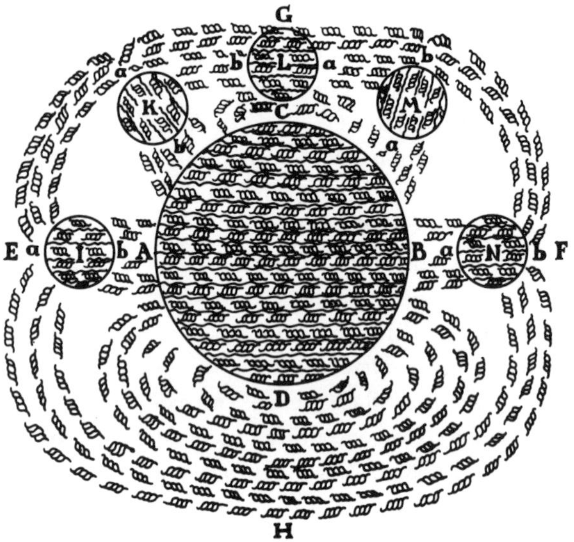

9. *Descartes's scheme explaining magnetic effects. Source: René Descartes,* Principles of Philosophy *(1644).*

flowing through appropriately sized pores to the brain and the organs of reproduction. The body's "animal spirits" were composed of the smallest and most highly agitated of the particles in the blood, which entered the cavities of the brain, then flowed through the hollow nerves and on into the muscles to produce sensory and motor effects that were explicable in the same terms used for artificial fountains and similar mechanical devices. So instances of what one would now term "reflex action" could be accounted for in appropriately, and specifically, mechanical terms. The particles of fire *A* (in fig. 10) move very quickly and therefore possess force sufficient to displace the adjacent skin *B;* this pulls the nervous thread *cc,* which opens the pore *de* that terminates in the brain, "just as, pulling on the end of a cord, one simultaneously rings a bell which hangs at the opposite end." That pore being duly opened, the animal spirits contained in brain cavity *F* enter and are carried through it, "part into the muscles that serve to withdraw this foot from the fire, part into those that serve to

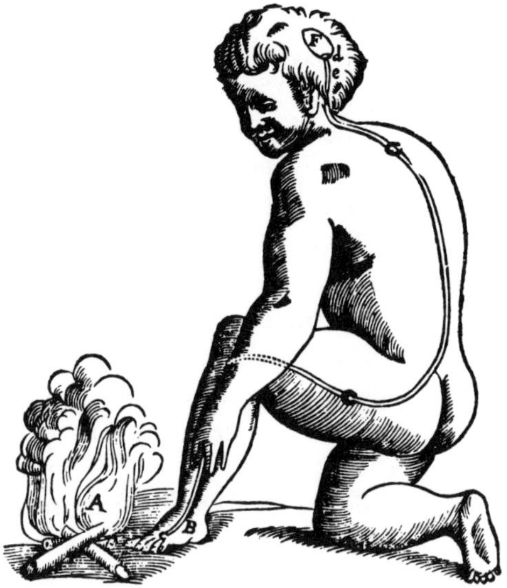

10. *Descartes's scheme explaining reflex action. Source: René Descartes,* Treatise of Man *(1664).*

turn the eyes and head to look at it, and part into those that serve to advance the hands and bend the whole body to protect it."

By contrast with Descartes's high degree of micromechanical specificity, English mechanical philosophers tended to follow Boyle in adopting a more cautious approach. Boyle was convinced that the original creation of the world had caused the division of a homogeneous "universal matter" into "little particles, of several sizes and shapes, variously moved." (It was for this reason that Boyle was content to designate the new philosophy either "mechanical" or "corpuscular.") These particles or corpuscles were then "associated into minute masses or clusters" differentiated by what Boyle referred to as their "textures," or the spatial arrangement of their parts. The qualities or properties of things were thus to be accounted for "by virtue of the motion, size, figure, and contrivance" of the corpuscles. Change in properties might be explained by changing the corpuscles' "texture" or states of motion. Where Boyle's practice diverged from Descartes's was in his extreme reticence in moving from mechanical principles to mechanical specifics, and the next chapter will show Boyle's diffidence in action in his account of such phenomena as air pressure. Matter and motion were, to be sure, the principles that were said to render mechanical accounts "intelligible"—one might model the invisible world of corpuscles on the visible and tangible phenomena presented by the behavior of medium-sized objects in the world of everyday experience.

Some philosophers conjectured that the newly invented microscope would soon make corpuscles visible: after all, did not instrumentally assisted sight already reveal macroscopically smooth surfaces to be microscopically rough (fig. 11)? The Dutch microscopist Antoni van Leeuwenhoek (1632–1723), who was loosely inspired by Descartes's theory of matter, initially reckoned that all bodies were composed of small "globules," the same globules he had repeatedly *seen* in a great range of microscopic observations. More reservedly, the English microscopist and experimentalist Robert Hooke expressed the hope that through improvements in the microscope we might *eventually* see "the figures of the compounding Particles of matter," and his colleague Robert Boyle even more cautiously concurred: "If we were sharp-sighted enough, or had such perfect microscopes, as I fear are more to be wished than hoped for, our promoted sense might discern . . . the particular sizes, shapes and situations of the extremely little bodies" that are, for example, the cause of color. Similarly, Hooke held out the possibility that the microscope might definitively take away the legitimacy of talk of "occult" qualities by making visible those "small Machines of Nature" by which effects are actually achieved. But most practitioners accepted that the corpuscular world was, and probably would forever remain, inaccessible to human vision and that micromechanical explanations of this form therefore necessarily had a *hypothetical* character—that is, their physical truth could never be *proved* by sensory means.

Corpuscularianism was advanced as a philosophically plausible way of making sense of the behavior of visible bodies, and it was rendered credible as microscopes revealed more and more qualitatively

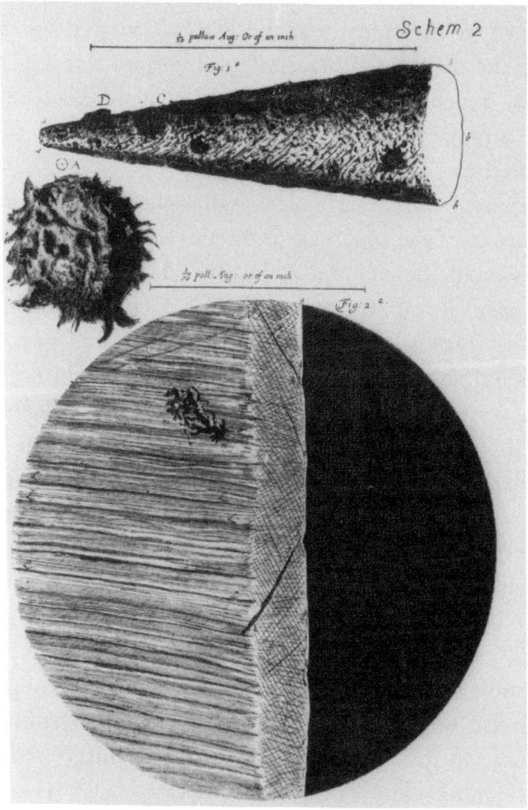

11. *Illustrations of microscopically enlarged common objects, from Robert Hooke's* Micrographia *(1665): at the top a needle point, below it a printed full stop or period, and at the bottom the edge of a sharp razor.*

different hidden appearances, and especially as more and more natural phenomena were shown to be compatible in principle with a matter-and-motion account. Like Descartes, Boyle wrote extensively in an endeavor to show how a great range of natural phenomena might be accounted for the by size, shape, texture, and motion of corpuscles. But unlike Descartes, Boyle rarely if ever spelled out what the relevant sizes, shapes, textures, and motions were that produced magnetism, cold, acidity, and the like. He viewed his task just as showing the power and plausibility of corpuscular explanations in principle. Seventeenth-century corpuscular mechanism therefore spanned a range from the methodologically general to the explanatorily specific.

Corpuscular and mechanical philosophers aimed to give a plausible account of the observed properties of bodies—their coldness, sweetness, color, flexibility, and the like—but they sought to do so by talking about a realm of corpuscles that could not be observed and that lacked those properties in themselves. So if one were to ask why a rose is red and sweet smelling, the answer would not be that its ultimate constituents possessed the properties of redness or sweetness. This point was of fundamental importance to the critique of Aristotelianism. On the one hand, the mechanical account was advertised as uniquely intelligible; on the other hand, it referred to an ultimate reality that had many properties qualitatively different from those available in common experience.

The distinction in question is customarily referred to as that between "primary" and "secondary" qualities, and though it became almost ubiquitous in seventeenth-century philosophy, no two versions of it were exactly alike. Although elements of that distinction appear in the work of such Greek "atomists" as Democritus (ca. 460–370 B.C.) and Epicurus (ca. 341–270 B.C.), its earliest clear seventeenth-century articulation is in Galileo's *The Assayer* of 1623. Here Galileo noted that people commonly have experience of objects they call hot. As a report of subjective sensation, there is nothing wrong with saying "this pot is hot." Where people go wrong, Galileo said, is in supposing that "heat is a real phenomenon, or property, or quality, which actually resides in the material by which we feel ourselves warmed." Although we cannot conceive of an object without thinking that it has a certain shape, size, and state of motion, Galileo noted that we can easily think of objects that are not red, or sweet, or hot. These latter qualities are what present themselves to our senses when we encounter a particular object, not what belongs to the object in itself: "Hence I think that tastes, odors, colors, and so on are no more than mere names so far as the object in which we place them is concerned, and that they reside only in the consciousness."

Primary qualities were those that *really belonged* to the object in itself: its parts' shape, size, and motion. They were called primary (or sometimes "absolute") because no object, or its constituents, could be described without reference to them. Secondary qualities—redness, sweetness, warmth, and so on—were *derived from* the state of an object's primary qualities. The primary caused (and was held to explain) the secondary. So, in a corpuscular philosophy, a body's constituent bits of matter were in themselves neither red nor sweet nor warm, but their size, shape, arrangement, and motions might produce these *subjective* effects in us. All the experienced diversity of natural objects was thus to be accounted for by the mechanically simple and primitive qualities that necessarily belonged to all bodies as bodies, and not to roses or iron bars or magnets as types of bodies. As the English philosopher John Locke (1632–1704) put it, "There is nothing like our ideas [of bodies], existing in the bodies themselves." The ideas we have of sweetness, redness, and warmth are but the effects *on us* of "the certain bulk, figure, and motion of the insensible parts" of bodies. Only *some* of our ideas of bodies might now be treated as objective—that is, corresponding to the nature of things themselves—and these would include our ideas of bodies as having certain shapes, sizes, and motions. However, other experiences and ideas would have now to be regarded as subjective—the result of how our sensory apparatus actively processes impressions deriving from the real, primary realm. Yet the rose of common experience is experienced not as an ordered aggregate of qualities but as itself: red, roughly circular, sweet smelling, three inches across, etc. The distinction between primary and secondary qualities, just like the Copernican view of the world, drove a wedge between the domain of philosophical legitimacy and that of common sense. Micromechanical reality took precedence over common experience, and subjective experience was severed from accounts of what objectively existed. Our actual sensory experience, we were instructed, offered no reliable guide to how the world *really was*. Accordingly, that fundamental distinction took a giant step, as the historian E. A. Burtt has written, toward "the reading of man quite out of the real and primary realm." Human beings, and human experience, were no longer to be taken as "the measure of all things."

In making this claim, the mechanical philosophers were setting themselves not just against common experience and common sense but also against the central Aristotelian doctrine of "substantial forms" (or "real qualities"). Medieval and early modern Aristotelians liked to make an analytic distinction between the "matter" and the "form" of bodies.[15] Loosely put, the matter of a marble statue is the material substratum out of which a statue of Alexander or a statue of his horse might be made. You can make a statue of anyone or anything at all out of marble, so the "matter" of a particular statue does not give an adequate account of what it is. The "form" of a given statue is that immaterial ordering principle that makes it a representation of Alexander or Alexander's horse. The "matter" out of which any given entity is made has no properties of its own; it is the "form" with which the "matter" is endowed that makes it this or that *kind* of body. Forms were real entities; they were not material but they were attached to matter. Similarly, one might speak of the substantial form of a rose or of a rat. The substantial form of these things was what gave the matter they contained its roseness or ratness. Any given rose or rat might have particular features that conferred its individuality, but these counted as "accidents" and had nothing to do with its substantial form—that is, with its being a rose or a rat. So for Aristotelians a physical account of things always had an irreducibly qualitative character: things were what they were, and not something else, because they had the real qualities of species within them. Our ordinary sensory perception of things was caused by the forms of

15. Strictly speaking, the doctrine of substantial forms rejected by seventeenth-century moderns was developed *from* Aristotle's writings by his Scholastic followers from the Middle Ages through the sixteenth and seventeenth centuries. Whether such a doctrine belonged properly to Aristotle himself is still a subject of scholarly debate.

things, and accordingly, there was a qualitative *match* between how the world was and how we experienced it.

These "substantial forms" were a favorite object of mechanical philosophers' ridicule, and the modern rejection of substantial forms helped mark out what it was to give a properly mechanical, and intelligible, account of nature. To Bacon, Aristotelian forms were "figments of the human mind." Boyle found it simply absurd to speak of forms as not material but as "belonging to" material bodies. These entities could not figure in proper physical explanations, and a matter-and-motion philosophy banished talk of such things. Substantial forms were identified as occult qualities. They were unintelligible, not part of a rightly constituted natural philosophy. Locke agreed that one could form no intelligible *idea* of immaterial substantial forms: "When I am told that [there is] something besides the figure, size, and posture of the solid parts of that body in its essence, something called *substantial form,* of that I confess I have no idea at all." For Hobbes all talk of "incorporeal substances" (including substantial forms) smacked of ideology. That such speech was central to Aristotelian natural philosophy was referred to its domination by priests, who used notions of substantial forms, separated essences, and incorporeal substances to grab a share of state power, to frighten the masses and keep them in awe. Material bodies just do not have forms or essences poured into them, as it were: their material nature—as defined by the mechanical philosophy—*is* their nature. What was not matter and not manifest from its effects was pronounced mysterious and occult, not intelligible, not belonging to the practice of a mechanical philosophy of nature.

Mechanical philosophers' reiterated insistence that their explanations were uniquely intelligible was therefore, as we have seen, a notable argument in their favor. One cannot understand how mechanical explanations were embraced, and how nonmechanical explanations were rejected, without appreciating the importance of this oft-asserted difference in intelligibility. Nevertheless, from a more disengaged point of view, there are certain problems worth noting about the basic structure, and therefore the scope, of mechanical accounts. Such explanations have a *structural* character. That is to say, the characteristics and behavior of a complex natural entity are to be explained by pointing to its composition—its constituent parts, their makeup, and their behavior. As we have seen, the structural explanations of the mechanical philosophy typically proceeded by way of "micromechanisms." So, for instance, one would explain heat by referring to the rapid and percussive motions of the invisible corpuscles of which hot bodies were composed. Or in an example to be treated in the next chapter, one would explain air pressure by pointing to the elastic characteristics of the invisible corpuscles that made up the air.

The intelligibility of such explanations flowed from the circumstance that in many cases one could point to *visible and tangible* exemplars from the everyday world of human life in which one could produce similar effects by mechanical means. It is a matter of common experience, and therefore readily intelligible, that we can produce heat by rapid and percussive motion of sticks or hands, and that we can keep ourselves from getting cold by putting our own bodies in rapid or percussive motion. (Here is another instance of the relation between the clarity of knowledge and the ability to construct the objects of knowledge that was noted above.) Yet in their more aggressive moods, mechanical philosophers sought to explain not just *some* natural phenomena but *all* of them. Descartes's *Principles of Philosophy* (1644) thus took on *everything* in nature—the gravitation of bodies, the behavior of liquids and magnets, the causes of earthquakes, chemical combination, the movements of human bodies, and the bases of human sensation, and so on and so on—and concluded by affirming that "there is no phenomenon in nature whose treatment has been omitted" and that cannot be accounted for by mechanical principles.

However, although micromechanical explanatory structures could readily be thought up for all natural phenomena, not all of them could draw on the intelligibility flowing from having mechanical counterparts in the realm of medium-sized objects that populate human experience. Take, for example, human sensation. Here Descartes notably offered extended mechanical explanations based on hydraulic principles and the mechanical operation of fluids, valves, and tubes—as in his account of bodily sensation of, and movement away from, the heat of a fire. But in the macroscopic domain there was nothing to explain how sensation was mechanically produced that enjoyed the intelligibility of, for example, a micromechanical kinetic explanation of heat or a micromechanical structural explanation of air pressure. For these reasons, some critically minded historians and philosophers have even wondered whether the claimed global intelligibility of mechanical explanations was more than just practitioners' *agreement* that such explanations would count as more intelligible than alternatives. When mechanical philosophers sought to explain pleasant and unpleasant smells or tastes by pointing to the rough or smooth texture of bodies' constituent particles, were they really offering something different from, and inherently more intelligible than, the explanations of their Aristotelian opponents? The historical philosopher Alan Gabbey thinks not: in the mechanical philosophy "the phenomena to be explained were caused by entities whose structures were such that they caused the phenomena. Previously, opium sent you to sleep because it had a particular dormitive quality: now it sent you to sleep because it had a particular corpuscular micro-structure that acted on your physiological structures in such a way that it sent you to sleep." From this perspective, the superior intelligibility, and therefore the explanatory power, of the mechanical philosophy was more limited than its proponents claimed. Adherents' *conviction* that mechanical accounts were globally superior to alternatives, and more intelligible, has to be explained in historical rather than abstractly philosophical terms.

The Mathematical Structure of Natural Reality

It is sometimes said that the mechanical picture of a matter-and-motion universe "implied" a mathematical conception of nature. Certainly a mechanical view of the world was in principle amenable to mathematization, and a number of mechanical philosophers vigorously insisted on the central role of mathematics in the understanding of nature. Boyle, for example, accepted that a natural world whose corpuscles were conceived to be variously sized, shaped, arranged, and moved called out, in principle, for mathematical treatment. Despite widespread contemporary professions of a natural "fit" between mechanism and mathematically framed accounts, however, very little of the mechanical philosophy was actually mathematized, and the ability to represent mathematically expressed physical regularities or laws did not depend on belief in their mechanical causes. That is to say, although the mathematization of natural philosophy was certainly an important feature of seventeenth-century practice, professions of a constitutive relation between mechanism and mathematics remain problematic.

Seventeenth-century confidence in the basic propriety and power of a mathematical framework for natural philosophy had ancient warrants. Modern natural philosophers turned to Pythagoras, and especially to Plato (ca. 427–347 B.C.), to legitimate a mathematical treatment of the world, quoting Plato's dictum that "the world was God's epistle written to mankind" and that "it was written in mathematical letters." Galileo argued that natural philosophy ought to be mathematical in form because nature was mathematical in structure. Modern natural philosophers, and not just those of the mechanical and corpuscular variety, were widely agreed that mathematics was the most certain form of knowledge, and for that reason one of the most highly valued. Yet the overarching questions for those concerned with the study of physical nature were how, in what ways, and to what extent it was proper to apply mathematical methods to the interpretation of real natural bodies and real physical processes. That it was *possible* to study nature mathematically was in principle not to be doubted, but was it practical and was it philosophically right to do so? Here there was important divergence of opinion among sixteenth- and seventeenth-century practitioners. Some influential philosophers were certain that the ends of science were, and ought to be, mathematically formulated binding laws of nature, while others doubted that mathematical representations could capture the contingencies and the complexities of real natural processes. Throughout the seventeenth century there were influential voices skeptical of the legitimacy of mathematical "idealizations" in the explication of physical nature as it actually was. Such practitioners as Bacon and Boyle said that mathematical accounts worked very well when nature was considered abstractly and less well when it was addressed in its concrete particularities. Galileo's mathematical law of fall pertained to ideal bodies moving in a frictionless environment. It is possible that no, or very few, real bodies have ever moved in precise obedience to such laws. Galileo announced that "motion is subject to the law of number," but the moving things concerned were only very approximately like the actual medium-sized bodies whose motions are the objects of daily experience. The question, to which chapter 2 will return, is whether natural philosophy was properly addressed to the domain of the mathematical ideal or to that of the concretely and particularly real, or whether some compromise position could be achieved.

Among the most full-blooded mathematical Platonists was Johannes Kepler, whose 1596 *Mysterium cosmographicum* ("The Secret of the Universe") announced a great discovery concerning the distances of the planets from the sun in a modified Copernican universe. Kepler's discovery was that the orbits of the six planets then known bore a striking resemblance to the distances from the sun that would be obtained if their "spheres" were inscribed within, and circumscribed by, the five regular solids of Plato's geometry: cube, tetrahedron, dodecahedron, icosahedron, octahedron (fig. 12). Inscribe a sphere within a great cube to represent the orbit of the outermost planet, Saturn. Nest within that the sphere of Jupiter inscribed within a tetrahedron, the sphere of Mars within that, and so on. Kepler's discovery was that the structure of the planetary system followed a geometrical order. And he offered a reason *why* it did so: "God, in creating the universe and regulating the order of the cosmos, had in view the five regular bodies of geometry as known since the days of Pythagoras and Plato, and . . . He has fixed, according to those di-

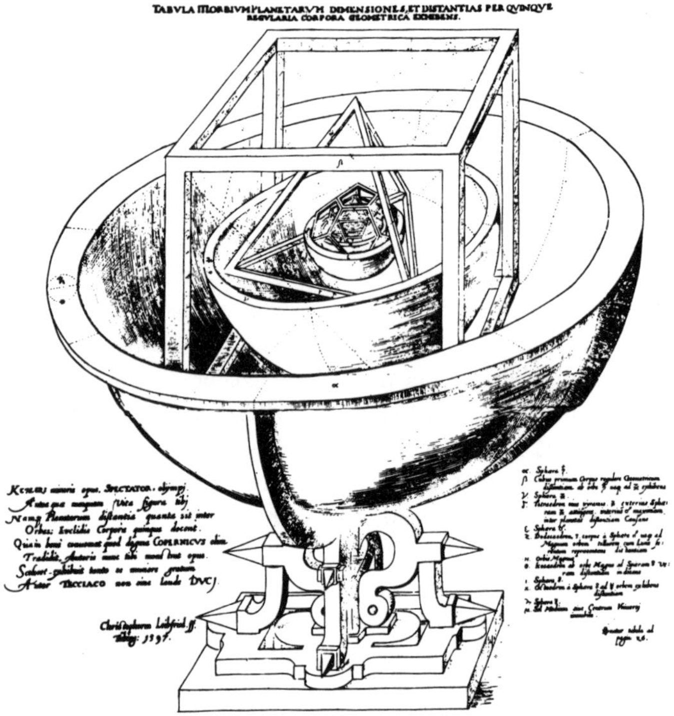

12. *Kepler's scheme showing the "Marvelous Proportion of the Celestial Spheres . . . established by means of the five regular Geometric solids." Source: Johannes Kepler,* Mysterium cosmographicum, *2d ed. (1621).*

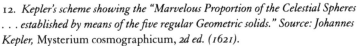

mensions, the number of heavens, their proportions, and the relations of their movements." A mathematically inclined astronomer had discovered that the creator God was a mathematician: the Creator had employed the principles of geometry to lay out planetary distances. The mathematical harmony of the spheres was a substantive feature of how the world was created and what principles governed its motions. Nature *obeys* mathematical laws because God had used these laws in creating nature.

The idea that nature obeys mathematical laws gave confidence to those promoting a mathematical conception of natural philosophy.

As investigators of physical phenomena, practitioners worked with, and tried to make sense of, real sensible, physical evidence; as mathematicians, they sought to establish the formal patterns that underlay, and may have given rise to, the natural world. This confidence reached its highest early modern development in the 1687 *Philosophiae naturalis principia mathematica* of Isaac Newton (1642–1727), the English title of which was *The Mathematical Principles of Natural Philosophy.* The world-machine followed laws that were mathematical in form and that could be expressed in the language of mathematics. Mathematics and mechanism were to be merged in a new definition of proper natural philosophy.

Newton's achievement was represented by many contemporaries as the perfection of the mechanical philosophy and by historians as the culmination of the Scientific Revolution. Certainly Newton decisively advanced the Galilean impulse to consolidate the domains to which a single natural philosophical scheme could be legitimately applied. The *Principia* unified mathematics with both celestial and terrestrial mechanics. Newton showed that the elliptical orbits of the planets previously described by Kepler were to be accounted for by two motions: one was inertial—planets tended to move with uniform velocity in a straight line, and therefore to fly off at a tangent to their orbits; the other was the *centripetal* gravitational attraction between planets and sun that tended to pull them toward the center of the solar system. All bodies whatever—celestial or terrestrial—tended to move uniformly in straight lines or to remain at rest; all bodies whatever—wherever they were—experienced gravitational attraction between each other. Gravitation is a universal force, acting in an inverse-square relation to the distances between bodies and describable by the mathematical equation $F = G(mm'/D^2)$. G is a constant, with the same value in all cases, no matter whether the force concerned acts between Mars and the sun, between Mars and Venus, or between this book in your hands and the earth below it. "All bodies whatsoever," Newton said, "are endowed with a principle of mutual gravitation."

The move toward the homogenization and the objectification of the natural world that was noted in Galileo's claims about sunspots at the beginning of this chapter was thus taken a giant step further. Historians have referred to Newton's achievement as the "destruction of the cosmos." Whereas traditional thought, and even much early modern thought, had conceived of a finite universe with qualitatively differentiated regions of space, Newton asserted an indefinitely sized universe united, as the historian Alexandre Koyré has said, "only by the identity of its fundamental contents and laws," a universe in which there is no qualitative physical distinction between heavens and earth, or any of their components, where "astronomy and physics become interdependent and united because of their common subjection to geometry." At the same time, proper knowledge of such a universe becomes itself objective. It is sometimes said that there was now no room for notions of purpose in this homogenized world "where abstract bodies move in an abstract space."[16] Only material causes exist in this abstract, homogenized world. All natural processes were now conceived to take place on a fabric of abstract time and space, self-contained, and without reference to local and bounded human experience. In the *Principia* Newton wrote down the definitions of terms necessary for the new practice: "Absolute, true, and mathematical time, of itself and from its own nature, flows equably without relation to anything external. . . . Absolute space, in its own nature, without relation to anything external, remains always similar and immovable." This new science was rendered perfect by creating for it a substratum divorced from the realms of the local, the bounded, and the subjective.

If there is wide agreement that Newton fulfilled the Galilean program, there was, and is, considerable divergence about whether Newton is rightly seen as perfecting a mechanical philosophy of nature. The gravitational force that bound the universe together was, to be sure, mathematically describable. It was even offered as a model

16. Chapter 3 will have to make some very significant qualifications to this sentiment, important as it was to traditional accounts of the identity of the Scientific Revolution.

for a practice whose end was the lawful characterization of the mathematical regularities of nature—laws (as Newton said) "deduced" from the actual observed behavior of bodies. The aim was physical *certainty,* and the tool for achieving that certainty was mathematics. Yet the price of that conception of science included at times a disengagement from inquiry into physical *causes.* So Newton freely acknowledged that "I have been unable to discover the cause of . . . gravity from phenomena, and I feign no hypotheses." He meant "only to give a mathematical notion of those forces, without considering their physical causes." The mathematization of the universe might then stand against the quest for causes, mechanical and material or otherwise. One interpretation of the Newtonian enterprise thus has it setting aside causal inquiry in favor of mathematical formulations of the regularities observable in nature, while another interpretation celebrates Newton's expansion of the scope of causal mechanical explanation.

Crucially, however, Newton reintroduced, or at least put new stress on the role of, immaterial "active powers" in a properly constituted natural philosophy, especially in accounting for effects whose reduction to mechanical principles he considered impossible or improper: magnetism, electricity, capillary action, cohesion, fermentation, and the phenomena of life. Although it might still be said that the preferred form of causal accounting was mechanical and material, in this version the practice of natural philosophy was no longer to be circumscribed by the provision of such accounts, and chapter 3 will treat the religious as well as the philosophical contexts that gave such a position much of its significance. Newton insisted that he had not sacrificed mechanism; such rival philosophers as the German Gottfried Wilhelm Leibniz (1646–1716) violently accused him of using the enormous cultural prestige of mathematics to reintroduce occult principles and of abandoning the dream of specifying a completely mechanical universe. For Leibniz, and others, the paramount condition of intelligibility was the provision of a plausible mechanical cause, and since Newton had not done so—as in the case of gravitation—his accounts were identified as unintelligible and occult. For Newton it was "absurd" to regard gravity as acting between bodies at a distance, without the mediation of material bodies, and he persistently tried to find a modus operandi for how gravitational attraction was conveyed through a medium. Yet even without that physical theory, gravitational attraction was *not* to be regarded as unintelligible: its intelligibility resided in the lawful account of its action. The law of gravitation could be used for explanatory ends even if no mechanical cause could be specified.

Accordingly, there can be no facile generalization about whether the Newtonian achievement should count as the culmination of the mechanical philosophy, as its subversion by the reintroduction of occult qualities, or as the creation of a new practice, to be judged by new philosophical standards. Late seventeenth- and early eighteenth-century philosophers debated just those points about the proper understanding of Newton's achievement. They disputed whether Newton had perfected mechanism or denied it; they debated whether mechanical causes had to be given as the condition for physical explanation. So too do historians, and so too do many present-day scientists.

Two

HOW WAS IT KNOWN?

Reading Nature's Book

Nothing so marked out the "new science" of the seventeenth century as its proponents' reiterated claims that it *was* new. Corpuscular and mechanical philosophers, on many occasions, vigorously insisted that their innovations represented radical departures from traditionally constituted bodies of natural knowledge. Text after text stipulated the novelty of its intellectual contents. In physics Galileo offered his *Discourses and Demonstrations concerning Two New Sciences;* in astronomy there was Kepler's *New Astronomy;* in chemistry and experimental philosophy Boyle published a long series of tracts called *New Experiments;* Pascal wrote about the vacuum in his *New Experiments about the Void,* as did Otto von Guericke in his *New Magdeburg Experiments on Empty Space.* Bacon's *New Organon* was labeled as a novel method meant to replace the traditional *organon* (Aristotle's body of logical writings), and his *New Atlantis* was an innovative blueprint for the formal social organization of scientific and technical research.

The very novelty of the emerging practices was often identified as a major point in their favor. Traditional stocks of knowledge, and traditional ways of securing and validating knowledge, were widely said to be worthless: they ought to be discredited and swept away. And in so doing the nature of "old" philosophies was often caricatured so as to misrepresent their complexity and sophistication. In seventeenth-century England self-styled "moderns" arrayed themselves against contemporary "ancients." The more polemical voices among the moderns reckoned that nothing ought to be preserved from traditional practices and that the textual legacy of ancient learning was little more than a testament to human capacity for delusion and human gullibility in being imposed on by authority. (In reply, the powerful community of seventeenth-century ancients identified their opponents as philistines who simply displayed their ignorance through their refusal to learn from the arduously compiled, and fundamentally sound, knowledge of the ancestors.)

Bacon's often-repeated architectural metaphor summed up the radical modernizing impulse. So worthless were traditional philosophies that "There [is] but one course left . . . to try the whole thing anew upon a better plan, and to commence a total reconstruction of sciences, arts, and all human knowledge, raised upon the proper foundations." In France, Descartes similarly announced that what then counted as philosophy had produced little of value. He shut himself up alone "in a stove-heated room" and set aside all the philosophic texts he had ever read. Starting the philosophic project all over would be better "than if I built only upon the old foundations." And the English experimentalist Henry Power (1623–68) precisely followed the pattern in applauding the new philosophy: "Me-thinks, I see how all the old Rubbish must be thrown away, and the rotten Buildings be overthrown. . . . These are the days that must lay a new Foundation of a more magnificent Philosophy, never to be overthrown." Indeed, the grip of this "modern" conception is reflected in dominant strands of historiography: twentieth-century historians and philosophers have only with difficulty sufficiently distanced themselves from the rhetoric of their predecessor moderns to offer a close assessment of the relation between seventeenth-century modernist rhetoric and historical realities.

Almost needless to say, no house is ever built of entirely virgin materials, according to a plan bearing no resemblance to old patterns, and no body of culture is able to wholly reject its past. Historical change is not like that, and most "revolutions" effect less sweeping changes than they advertise or than are advertised for them.[1] Copernicus's new astronomy preserved Aristotle's assumption about the perfection of circular motion, as did William Harvey's (1578–1657) discovery of the circulation of the blood. The very identity and practice of early modern astronomy depended utterly on the observational data compiled by the ancients: there was no way that sixteenth- and seventeenth-century practitioners, however "revolutionary" minded, could set aside that legacy. As chapter 3 will show, many mechanical philosophers publicly announced their rejection of the old teleology while preserving an important role for explanations in terms of purpose in some of their actual explanatory practices. Modernist rhetoric embracing the totally new and wholly rejecting the past does not adequately describe historical reality. Copernicus himself, and many of his followers, liked to argue that heliocentrism was in fact an ancient view, corrupted out of recognition by subsequent accretions, and the Flemish anatomist Andreas Vesalius (1514–64)—celebrated as the inventor of rigorous observational methods and as the critic of ancient anatomical claims—saw himself as reviving the pure medical knowledge of the Greek physician Galen (A.D. 129–ca. 200). And even as Cartesian methodology was celebrated for its radical replacement of existing knowledge-making practices, so its author was seen by some contemporaries to occupy a traditional role as

1. Indeed, Descartes explicitly worried about the effects his version of methodological modernism would have were it to be collectively adopted: "There is no plausibility in the claim of any private individual to reform a state by altering everything, and by overturning it throughout, in order to set it right again. Nor is it likewise probable that the whole body of the Sciences, or the order of teaching established by the Schools, should be reformed. . . . In the case of great bodies it is too difficult a task to raise them again when they are once thrown down . . . and their fall cannot be otherwise than very violent." His maxims of method were intended, so he said, *for himself,* given his own situation and his own idiosyncratic temperament, although one may well speculate whether that restrictive caution was ingenuous or likely to be effective.

great philosophical master: "Behold! he has become the New Aristotle."

"New" and "old" views of nature coexisted, their adherents occasionally contesting for the right to be regarded as modern or ancient. Some practitioners asserted the primitive antiquity of what was apparently new, while others argued that what seemed traditional was actually up-to-date and intellectually unsurpassed. Chapter 1 referred to Bacon's view that modern advances in natural knowledge were the fulfillment of Old Testament prophecy, and chapter 3 will note that several modern practitioners conceived of increasing technical control within a Christian messianic context. For every practitioner who equated the innovatory with the valuable there was another who linked modern opinions with uneducated ignorance. The Scientific Revolution was significantly, but only partially, a New Thing. Nevertheless, the rhetoric of wholesale rejection and replacement draws our attention to how practitioners tended to position themselves with respect to existing philosophical traditions and institutions.

What was said to be overwhelmingly wrong with existing natural philosophical traditions was that they proceeded not from the evidence of natural reality but from human textual authority. If one wished to secure truth about the natural world, one ought to consult not the authority of books but the authority of individual reason and the evidence of natural reality. The English natural philosopher William Gilbert (1544–1603), for example, dedicated his 1600 book on magnetism to "true philosophers, ingenuous minds, who not only in books but in things themselves look for knowledge." This was, Gilbert said, "a new style of philosophizing." When Descartes shut himself up alone, it was an expression of a resolve "to seek no other knowledge than that which I might find within myself, or perhaps in the great book of nature." And William Harvey said that it was "base" to "receive instructions from others' comments without examination of the objects themselves, [especially] as the book of Nature lies so open and is so easy of consultation." Here was one of the central rhetorical figures that the new philosophical practitioners used to distinguish themselves from the old. The proper object of natural philosophical examination was not the traditionally valued books of human authors but the Book of Nature.

The Swiss Renaissance medical man and natural magician Paracelsus (1493–1541) argued vehemently that those who sought medical truth should put aside the ancient texts and take themselves directly to the study of herbs, minerals, and stars. Natural reality is "like a letter that has been sent to us from a hundred miles off, and in which the writer's mind speaks to us." He said that he did not "compile his textbooks from excerpts of Hippocrates and Galen" but wrote them anew, "founding them upon experience": "If I want to prove anything I do so not by quoting authorities but by experiments and reasoning." When Galileo advocated a mathematically conceived natural philosophy, he used the figure of nature's book to argue his case: "Philosophy is written in this grand book, the universe, which stands continually open to our gaze. . . . It is written in the language of mathematics, and its characters are triangles, circles, and other geometric figures without which it is humanly impossible to understand a single word of it." In the 1660s Boyle wrote that "each page in the great volume of nature is full of real hieroglyphs, where (by an inverted way of expression) things stand for words, and their qualities for letters." Few modern natural philosophers omitted to make reference to the Book of Nature, recommending its direct inspection over the texts of human authorities, however ancient and however highly valued they had been.

No seventeenth-century modernist maxims seem more self-evidently sound than these: rely not on the testimony of humans but on the testimony of nature; favor things over words as sources of knowledge; prefer the evidence of your own eyes and your own reason to what others tell you. Here is the root idea of modern *empiricism,* the view that proper knowledge is and ought to be derived from direct sense experience. And here too are the foundations of modern mistrust of the social aspects of knowledge making: if you really want to secure truth about the natural world, forget tradition, ignore authority, be skeptical of what others say, and wander the fields alone

13. *Johannes Hevelius and his second wife, Elisabetha Koopman, making astronomical observations with a sextant. Elisabetha—thirty-six years younger than her husband and bringing to the marriage a good fortune—played a valuable part in financing the observatory and in making and recording astronomical observations, and after Hevelius's death she arranged for the publication of works from his papers. Source: Johannes Hevelius,* Machina coelestis pars prior *(1673).*

14. *An astronomer, probably Hevelius himself, making a telescopic observation. Hevelius was regarded as having particularly keen and penetrating eyes, and his observations were in general viewed as highly accurate and reliable. His observatory was on the roof of his own house, and by the 1660s it was one of the premier observatories in Europe. Source: Johannes Hevelius,* Selenographia *(1647).*

with your eyes open. As John Locke said, "We may as rationally hope to see with other men's eyes, as to know by other men's understandings. So much as we ourselves consider and comprehend of truth and reason, so much we possess of real and true knowledge. . . . In the sciences, every one has so much as he really knows and comprehends. What he believes only, and takes on trust, are but shreds."

There is probably no other sensibility that more strongly links seventeenth-century and late twentieth-century moderns than the recommendation of intellectual individualism and the rejection of trust and authority in the pursuit of natural knowledge. Yet the rhetoric of individualistic empiricism was neither unalloyed nor self-evident for early modern practitioners. Both the practice of observation and the credibility of observation reports in the early modern period could be intensely problematic. Triumphalist histories of science sardonically relate the story of the foolish professor from Padua who refused to look through Galileo's telescope to see with his own eyes the newly discovered moons around Jupiter. What can one say about a man who prefers the authoritative tradition maintaining that such moons could not exist to the evidence of his own eyes? For twentieth-century moderns, merely to describe such behavior is to condemn it as absurd.

Nevertheless, in early modern culture there were well-entrenched justifications for such an apparently bizarre preference. If, for example, Galileo's telescopic observations were to count as evidence for or against astronomical theories, there had to be grounds for assurance that this evidence was secure. Such assurance was practically available for telescopic observations of *terrestrial* things. When Galileo went to Rome in 1611 to demonstrate his telescope, he gathered a number of eminent philosophers on top of one of the city gates. Peering through the telescope from this vantage point, they were able to see the palace of a nobleman "so distinctly that we readily counted its each and every window, even the smallest; and the distance is sixteen Italian miles." And from the same point observers could read the letters on a gallery distant two miles, "so clearly, that we distinguished even the periods carved between the letters." So the reliability of telescopic observation of terrestrial objects could be vouched for by comparing what was seen through the instrument to what was known by unmediated inspection.

However, in early modern philosophical culture matters might be otherwise when the telescope was directed at the heavens. Both formal and informal contemporary theories of human vision might count against the reliability of telescopic observation of the heavens. We are familiar with terrestrial objects and their backgrounds, and we use that familiarity automatically to correct a range of apparent instrumental distortions. That same familiarity is not available for observations of heavenly bodies. To grant the reliability of Galileo's telescope in providing evidence of the heavens therefore might involve a call for a new and powerful theory of vision, and there is little evidence that Galileo had such a theory.

Even if, unlike the Paduan professor, one took up Galileo's offer and looked directly through the telescope at heavenly bodies, there was no guarantee that one would see what Galileo said he saw. Following his discovery of the moons of Jupiter, Galileo on several occasions assembled eminent practitioners to witness them. Many of these witnesses allowed that though the telescope worked "wonderfully" for terrestrial vision, it failed or "deceived" in the celestial realm. One witness wrote that Galileo "has achieved nothing, for more than twenty learned men were present; yet nobody has seen the new [moons] distinctly. . . . Only some with sharp vision were convinced to some extent." Nor should such a state of affairs come as a great surprise. Seeing with the aid of a telescope (or a microscope) is skilled seeing under special conditions. When you and I learned these skills as students, we had enormous advantages over Galileo's contemporaries. We belonged to a culture that had already granted the reliability of these instruments (properly used), that had already decided for us what sorts of things authentically existed in the domains of the very distant and very small, and that had provided structures of authority within which we could learn what to see (and what to disregard). None of these resources was unproblematically available to Galileo; they had to be laboriously created and disseminated. So although it is right to say that instrumentally mediated experience of the heavens figured importantly in the evaluation of astronomical theories, it is vital to understand how precarious such experience might be and how much work was required to constitute it as reliable.

And there were still other general problems attending the modern use of individual sensory experience to evaluate traditionally established bodies of knowledge. Christian theology assured the devout that the senses of human beings following the fall from grace were corrupt, and that reliable knowledge was not to be had by trusting such debased sources. Among the moderns Bacon was far from alone in wholly accepting that before the Fall Adam had possessed "pure and uncorrupted natural knowledge," the power that allowed him to give creatures their proper names. Galileo maintained that Solomon and Moses "knew the constitution of the universe perfectly," and later Boyle and Newton reckoned that there might be a chain of specially endowed individuals through whom the pure and powerful ancient wisdom had been handed down intact, both intimating that they themselves might be present-day members of this lineage. In a more secular idiom, the idea of linear, cumulative intellectual progress was still novel and not widely accepted. Many scholars, including some of the more prominent natural philosophers of the early modern period, accepted as a matter of course that the ancients had better knowledge, and more potent technology, than that possessed in the sixteenth and seventeenth centuries or than any that modern human beings could have. The ruins of still-unsurpassed Greek and Roman engineering works appeared to strongly support that idea. Moreover, the "self-evident" view that testimony and authority are to be resorted to only when we cannot have individual experiential access is, as the philosopher Ian Hacking has noted, a creation of the very sixteenth- and seventeenth-century culture whose judgments we want to understand: "The Renaissance had it the other way about. Testimony and authority were primary, and things could count as evidence only insofar as they resembled the witness of observers and the authority of books."

One strong argument in favor of trusting to personal observation rather than traditional texts was more ancient than modern in flavor.

One might accept that truly ancient texts were enormously valuable sources of truth about the natural world while claiming that the originally pure sources of that ancient truth had been polluted over time. Robert Hooke drew an analogy between political and philosophical decay: "It is said of great Empires, that the best way to preserve them from decay, is to bring them back to the first Principles . . . on which they did begin. The same is undoubtedly true in Philosophy." One was to get ahead by going back: progress through purification. At one extreme Bacon accounted Aristotle himself a corrupter of a more primitive and valuable natural philosophy, while other moderns paid homage to Aristotle but not to early modern Aristotelians. William Gilbert was one of many modern natural philosophers who rejected the authority of the Schools by impugning the pedigree that seemed to link their doctrines to their alleged pure, potent, and ancient sources: the errors cumulatively introduced into ancient texts by "mere copyists" must be corrected by insisting on direct experience. Bacon suggested that the sciences "flourish most in the hands of the first author, and afterwards degenerate." The very antiquity of the Aristotelian tradition meant that the original body of knowledge had passed through many hands, each of which might have introduced corruption: "Time is like a river which has brought down to us things light and puffed up, while those that are weighty and solid have sunk." And in a similar vein, Boyle wrote that he did not trust every writer's quotations from other authors: many times, on inspection, the quotation was incorrect, and sometimes it had been wilfully fabricated.

The tradition of Renaissance scholarship known as humanism figured importantly in the complex relations between the value placed on individual experience and the authority of ancient texts. Humanism was a cultural practice that aimed at reforming the public stock of knowledge by close scholarly reinspection of the original Greek and Latin sources—setting aside later commentaries (and possible corruptions) by Christian and Arab writers. Humanist scholars suspected that ancient truth had been eroded by copyists and commentators over the centuries, and the practice of humanistic literary scholarship commonly was closely joined to that of observational science. So, for example, while some sixteenth- and seventeenth-century astronomers, including Kepler (fig. 15), insisted on the relative crudity of ancient astronomy compared with its modern perfection, others, including Newton, saw their task as including the recovery of the lost wisdom of the ancients, undertaking painstaking philological studies to support that enterprise. Some humanists even concluded that pristine textual truth could now be restored only by engaging the evidence of nature directly. Individual observation might therefore be a means to decide which copies of Greek and Latin manuscripts were indeed authentic.

This subtle and consequential humanist spur to direct observation was a feature of a wide range of early modern scientific practices, but it was perhaps most striking in sixteenth-century natural history. Here it was understood that currently available texts by such Greek and Roman natural historians as Theophrastus (ca. 372–287 B.C.), Pliny the Elder (A.D. 23–79), Dioscorides (fl. A.D. 54–68), and Galen were problematic copies of copies. Editions were known to vary, to be incomplete and corrupt. How could one discover the original pure and accurate descriptions of plants and animals? Where human history was concerned, the only major methods available were philology and the collation of text with text. For botany and zoology, however, one might also importantly collate texts with *the directly observed appearance of the living things in question*—on the culturally innocuous assumption that the forms of plants and animals had not changed over the intervening years. Observation could help decide what the original ancient descriptions had actually been and, further, what ancient names and descriptions referred to what existing plants. After all, wasn't this what the ancient authorities themselves had done? Hadn't Aristotle been a close observer of the natural world? Hadn't Galen advised practitioners "to become expert in all matters of plants, animals, and metals . . . by personally inspecting them, not once or twice, but often"? And when direct observation had been accomplished, sixteenth-century botanists went their ancient sources one better. The ancients had not supplemented their verbal descriptions

15. *The* Astro-poecilo-pyrgium *(a "variegated star tower" or "temple of astronomy") from the title page of Johannes Kepler's* Tabulae Rudolphinae *(1627). Kepler here gave iconographical expression to his belief in the genuine progress of astronomy from antiquity to the seventeenth century. The more rough-hewn architectural orders (at the rear) represent the crudity of ancient astronomy, while Copernicus and Tycho Brahe are placed by elegant Ionic and Corinthian columns bearing their names. The moderns are thus identified as older than the ancients, that is, as more sophisticated and knowing. The map at the front on the base shows the Danish island of Hven, where Tycho Brahe's observatory was situated, and the seated figure to the left is Kepler himself.*

with pictures—worrying that the human artist could not copy any given plant with the requisite accuracy or capture the seasonal variations in their appearance—but the printed books of such Renaissance German botanists as Otto Brunfels (ca. 1488–1534) and Leonhard Fuchs (1501–66) offered detailed woodcut illustrations to act both as standing records of botanical reality and as gauges to discipline the observations of others (fig. 16).

Christian religious impulses also figured in this connection. The Book of Nature that one was enjoined to read in preference to the texts of the Schoolmen was understood to be divinely written. It was widely said that God had written two books by which his existence, attributes, and intentions might be known. The one was Holy Scripture, but the other was increasingly referred to in the early modern period as the Book of Nature.[2] The Protestant Reformation of the sixteenth century laid special stress on the desirability of each Christian's having direct engagement with Scripture, not relying on the interpretations of priests and popes, and the invention of printing with movable type in the 1450s made the injunction to read the Bible for oneself more practically realizable. A similar impulse informed the encouragement to read the Book of Nature for oneself, not relying on the traditional interpretations of institutionalized authority. Direct experience of nature was accounted valuable insofar as it was understood to be engagement with a divinely written text.

Of course, more characteristically modern sentiments also figured in the preference for the evidence of things over the authority of texts. Chapter 1 noted opinions that the natural world to which modern philosophers enjoyed access was simply much larger and more varied than that known by the ancients. Philosophical schemes based on restricted knowledge were likely to be faulty for just that reason, and the expanded experience afforded, for example, by the voyages of discovery to the New World was an important support for cur-

2. The metaphor of nature's book was present in the early Christian period, Saint Augustine notably giving voice to it in the late fourth century. However, it received new emphasis and point in the Renaissance and early modern period.

PICTORES OPERIS,

Heinricus Füllmaurer. Albertus Meyer.

SCVLPTOR

Vitus Rodolph. Speckle.

16. *Making botanical observations and representations in the mid-sixteenth century. This illustration is taken from Leonhard Fuchs's 1542* De historia stirpium *("On the History of Plants"). It shows the different roles of the individuals who drew the specimen (here a corn cockle) from its natural appearance (*top right*), who transformed the drawings to woodblocks (*top left*), and who did the cutting of the final blocks (*bottom*). Fuchs, as author of the text, assured readers that "we have devoted the greatest diligence to make sure that every plant is depicted with its own roots, stalks, leaves, flowers, seeds, and fruits."*

rents of early modern skepticism about traditional philosophical systems. The Renaissance witnessed a revaluation of the possible scope of human knowledge and greater confidence about the prospects for intellectual and technical progress. Aggressive seventeenth-century modern rhetoric simply rejected traditional deference to ancient doctrine and anciently established stocks of knowledge. It began to be possible to denigrate not just the modern disciples of the Greeks but Aristotle himself. Few moderns followed Hobbes in describing Greek philosophy as full of "fraud and filth," but many found ways of asserting its inadequacy compared with modern modes of reasoning and direct engagement with nature itself. It was only through changed values placed on direct experience and textual authority that the former could possibly be taken to refute the latter.

Modern philosophers like Bacon and Hobbes simply inverted the historical scheme that gave antiquity its intellectual authority. "The old age of the world," Bacon wrote, "is to be accounted the true antiquity; and this is the attribute of our own times, not of the earlier age of the world in which the ancients lived." There is no justification for deference to the Greeks; it is we ourselves who enjoy the cumulative benefit of experience and the wisdom this produces: "With regard to authority, it shows a feeble mind to grant so much to authors and yet deny time his rights, who is the author of authors, nay, rather of all authority. For rightly is truth called the daughter of time, not of authority." Simply by virtue of being modern we know more and better than the ancients. The idea of intellectual progress was thus rendered historically natural.

The Constitution of Experience

In principle, therefore, the moderns' recommendation was clear: obtain experience yourself; mind not words nor traditional authority but things. Experience was to be formed into the foundations of proper scientific knowledge, and experience was to discipline theorizing about how nature in general worked. But what *kind* of experience was to be sought? How was it to be reliably attained? And how ought one to *infer* from experience to generalizations about the natural order, on whatever scale? Modern natural philosophical practices diverged importantly on these questions. What counted in one practice as reliably constituted experience, and reliable inference, was commonly identified by another as insecure or unphilosophical. One strand of philosophical practice was arguably continuous with Aristotle, although it appealed to some seventeenth-century moderns as well as contemporary ancients. The goal here was typically scientific *demonstration,* that is, the exercise of showing how conclusions about natural effects followed necessarily from indubitable, rationally established knowledge of their causes. For abstract mathematical sciences the principles one was to proceed from were taken as evident and indisputable, like an axiom in Euclid's geometry: "the whole is greater than any of its parts." For those sciences dealing with the physical world, principles rested on empirical claims that were reckoned to have something of the self-evident character of mathematical axioms.

The historian Peter Dear has noted that the term "experience" in sixteenth- and seventeenth-century Scholastic natural philosophy referred to a universal statement of fact. As such, it was supposed to be developed from reliable recollection of very many accessible instances, and its universality marked its status as an indubitable truth that could form a premise in logical scientific demonstration of the causal structure of the world. For Aristotle and many of his followers, natural phenomena were taken as givens: they were statements about how things behave in the natural world, and they might be derived from any number of sources—common or expert opinion, what was available to be sensed by competent persons or hauled up out of their memories. Take, for example, the experience—cited by Aristotle and his followers as evidence that the earth did not move—that an arrow shot straight up in the air landed about where it started. Or take the experience that heavy bodies fall, or that the sun sets in the west. These are all authentic experiences by virtue of appealing to "what any competent person knows," and they counted as statements about how the world in general behaves, as Aristotle said, "always or for the most part." The experience that figured in this practice rarely was specially laid on or laboriously obtained *particular experience*—its accessible and commonsensical quality was just what was required for this strand of natural philosophy. Experience had a significant role in this connection, but it was subordinated to an overall argumentative structure aimed at securing natural knowledge of a general and an indubitable character. To arrive at philosophically certain knowledge of the ordinary course of nature one had to reason upon experiences that reliably testified to that ordinary course. Indisputable and global conclusions required indisputable and global premises. Discrete and particular events might not answer to that purpose, and knowledge of them might be unreliable: the testifying person might be lying or deluded; the instruments used might distort the natural order of things; the events reported might be not ordinary but anomalous.

This preference for experience as "what happens in the world" formed an important basis for the practice of such moderns as Galileo, Pascal, Descartes, and Hobbes. And though these moderns criticized many of the claims, concepts, and procedures of Aristotelian natural philosophy, the "experience" they appealed to was often construed in a recognizably traditional way. So Descartes remarked that though experiments were necessary in natural philosophy, it was generally "better to make use simply of those which present themselves spontaneously to our senses," and Hobbes reckoned that artificial experiment was unnecessary since there was sufficient experience "shown by the high heavens and the seas and the broad Earth."

Some of the most celebrated early experimental performances of the Scientific Revolution were Galileo's "inclined-plane experiments," in which balls were rolled down a smooth ramp to provide the empirical warrant for the mathematically expressed law of fall. Galileo's report of these experiments—which he enlisted against the Aristotelian account of motion—announced that he had done the experiments "often," even as often as "a hundred times," the results fully agreeing with his theory. In fact, historians have long debated

17. *Depiction of a hydrostatic experiment reported by Pascal in 1663. In the following year Boyle expressed pointed skepticism that Pascal "actually made" the experiment in question. Perhaps, Boyle said, Pascal "thought he might safely set [this result] down, it being very consequent to those principles of whose truth he was already persuaded." Such "thought experiments" did not, in Boyle's view, belong to a proper natural philosophy. Source: Blaise Pascal,* Traitez de l'équilibre des liqueurs *(1663).*

whether these experiments were ever actually performed or whether they are best regarded as "thought experiments," imaginative rehearsals in Galileo's mind of what *would* happen were certain manipulations to be carried out, given what we already securely know about the physical world.[3] Here, as Peter Dear has pointed out, Galileo was not saying, "'I did this and this, and this is what happened, from which we can conclude . . .'" Rather, he was saying, "'This is what *happens.*'" The experience that was produced and made public as a result of communicating this experiment was therefore *like* the experience that would result from doing the experiment imaginatively *in your head,* whether or not Galileo physically performed the particular experimental manipulations in question.

This kind of attitude toward experience as "what happens in nature" continued to be an important feature of *both* much modern and much Aristotelian practice on the Continent. It was a major task of Continental Jesuit practitioners to come to terms with the new stock of particular, and artificially obtained, experience afforded, for example, by the telescope and barometer, and to bring such findings within the compass of Aristotelian conceptions of the proper role of experience in philosophizing. This they did by deploying a wide range of social and linguistic techniques to give such particular experience the aura of certainty that Aristotelian philosophical practice deemed necessary, including the naming of reliable witnesses, the public display of relevant expertise, and the use of narrative techniques designed to make empirical statements look like indubitable axioms. It is simply not true, therefore, that seventeenth-century Aristotelian natural philosophy lacked the resources to come to terms with the new experience afforded by artificial experiment or scientific instruments, nor did Aristotelian frameworks immediately wither away with the appearance of modern alternatives. Right through the seventeenth cen-

3. Similar doubts about actual performance have attached to Pascal's Puy de Dôme experiment recounted in the preceding chapter and to one of his hydrostatic experiments that required a person to sit twenty feet underwater holding a cupping glass on his thigh over a long period (fig. 17).

tury traditions of Aristotelian natural philosophy, with their associated conceptions of experience, remained vigorous.

Yet many other seventeenth-century practitioners, especially in England, developed a new and quite different approach to experience and its proper role in natural philosophy. Early in the century Bacon influentially argued that the condition for a proper natural philosophy was its foundation in a laboriously compiled factual register of natural history—a catalog, compilation, and collation of all the effects that could be observed in nature. The register of natural historical facts was to contain several kinds: naturally occurring entities and effects, whether they were produced in the ordinary course of nature or were nature's "errors" or "monsters" (fig. 18), and those that might be artificially produced by human labor—"when by art and the hand of man she is forced out of her natural state, and squeezed and moulded"—that is, when nature is put to experimental trial or subjected to technological intervention. *First* natural history (the reformed and purified register of effects), *then* natural philosophy (reliable knowledge of the causal structure of nature that produced such effects). And a central feature of a closely related strand of modern natural philosophical practice was that it relied for its empirical content not just on naturally available *experience* of what went on in the world but also on *experiments* artificially and purposefully contrived to produce phenomena that might not be observed, or at least not easily, in the normal course of nature. These experiments typically involved the construction and use of special apparatus, such as the barometer described in chapter 1. The barometer, remember, was an instrument that was advertised to make the weight of the air—which we ordinarily do not experience—easily sensible, even visible.

The Control of Experience

Natural philosophy had gone spectacularly wrong, in Bacon's view, because it had been inadequately informed about the entities and

18. *A monstrous (or "freak") rooster, with a "quadruped's tail and a chicken's crest," observed by the Italian naturalist Ulisse Aldrovandi (1522–1605). Aldrovandi specified that he saw this chicken himself "when it was alive, in the palace of the Most Serene Grand Duke Francesco Medici of Tuscany; it struck fear into brave men with its terrifying aspect." Such monsters were often taken as divine omens and portents, and images of them circulated widely in early modern Europe. Source: Ulisse Aldrovandi,* Ornithology *(1600).*

phenomena that nature actually contained. Hitherto, Bacon claimed, "no search has been made to collect a store of particular observations sufficient either in number, or in kind, or in certainty, to inform the understanding." Gesturing at the illustrative role of experience described in the preceding section, Bacon judged that the ills of contemporary natural philosophy had arisen from an impoverished, and inadequately evaluated, stock of experience: "Just as if some kingdom or state were to direct its counsels and affairs not by letters and reports from ambassadors and trustworthy messengers, but by the gossip of the streets; such exactly is the system of management introduced into philosophy with relation to experience. Nothing duly investigated, nothing verified, nothing counted, weighed, or measured, is to be found in natural history; and what in observation is loose and vague, is in information deceptive and treacherous." If experience were to form the foundations of a true and a useful philosophy of nature, in Bacon's view, it had to be *genuine,* actually occurring, specific experience. The point of such an exercise was not to use indubitable experience to *illustrate* a general view of how nature worked but to assemble enough authentic experience to ground inquiries into how nature might plausibly work.

In this connection, some further important qualifications now have to be made to modernist rhetoric about the respective roles of experience and authority noted earlier in this chapter. Modernist rejection of authority and testimony in science was in fact finely focused. When seventeenth-century moderns recommended doing away with authority, they generally had in view the traditionally established authority of Aristotle and his Scholastic followers in the universities. But despite much rhetoric preferring the authoritative testimony of things to that of people, the modern enterprise *in no way* dispensed with reliance on human testimony, nor is it possible to imagine what a natural scientific enterprise that wholly rejected testimony would look like. Modern practitioners were supposed to acquire a stock of factual knowledge, but most of that knowledge was necessarily acquired at second hand, and the treatment of Boyle's experimental writing later in this chapter will describe some techniques that aimed to reliably extend experience by indirect means.

The practical task, therefore, was to sift and evaluate experience reports. The poor state of existing natural philosophy was widely ascribed to inadequate quality control over its register of facts. If just any experience report were to be credited, then the house of natural philosophy would resemble the Bedlam or Babel that some moderns suggested it was. Bacon proposed a set of techniques for establishing that facts about nature were adequately observed, testified to, and recorded. Nothing was to be admitted into the register of natural fact "but on the faith of eyes" (that is, by eyewitness) "or at least of careful and severe examination": "away with antiquities, and citations or testimonies of authors"; "all superstitious stories" and "old wives' fables" were to be set aside (fig. 19). Indeed, William Gilbert's rhetorical dismissal of traditional natural historical testimony by comparing it to "the maunderings of a babbling hag" was echoed by many other moderns. If authorial testimony had to be used, then that circumstance ought to be carefully noted, along with the likely trustworthiness of the source: "Whatever is admitted must be drawn from grave and credible history and trustworthy reports." So experience was to be welcomed by this reformed natural philosophy as a powerful means of supplanting traditional practice, but experience reports had to be carefully monitored to ensure that they were genuine. The house of natural philosophy was indeed to be opened up, but entrance to its interior rooms was to be vigilantly controlled. And though several modern practitioners proffered explicit rules for evaluating experience reports, it needs to be stressed that formal methodology was far less relevant here than the mobilization of everyday social knowledge. Most practitioners seemed to *know* the visible signs of a trustworthy report and a trustworthy person without having the grounds of trustworthiness formally spelled out.[4]

4. Some recent historical work has pointed to the importantly *gentlemanly* constitution of the new practice, and consequently to the importance of gentle codes of honor and truth telling. It is quite possible that many practical problems of scien-

19. *A late sixteenth-century representation of acephalous American Indians. The belief that distant parts of the world were inhabited by strange peoples who "have no heads" and whose "eyes be in their shoulders" was current in antiquity, and it was given new life by European travelers (or alleged travelers) in the fourteenth century. The travel tales of Sir John Mandeville in the 1370s located such marvelously formed people in the East Indies, and in 1604 Shakespeare's Othello astounded Desdemona by telling her "of the Cannibals that each other eat, / The Anthropophagi, and men whose heads / Do grow beneath their shoulders." For those whose purpose was the reform of natural history, such testimony became emblematic of the problem of sorting out the genuine from the fabulous. Source: Levinius Hulsius,* Kurtze wunderbare Beschreibung *(1599).*

The Mechanics of Fact Making

The experiential facts providing the foundations of a reformed natural philosophy were to be statements not of "what happens in nature"

---

tific credibility were solved by a device as apparently simple as the gentlemanly code of honor, though considerations of technical *expertise* and *plausibility* were also undoubtedly important in the evaluation of many experience reports.

but of "what actually *happened* in nature" when observed in specific ways and in specific times, places, and circumstances, by specific people. For many natural philosophers, especially but not exclusively in England, this particularity was what made experience sufficiently reliable to *found* philosophical inquiry. Just because the register of fact was meant to provide the secure foundations of natural philosophy, the facts concerned must not be idealized or colored by theoretical expectations but must be ascertained and represented exactly as they presented themselves: not, for example, how stones fall, but how *this* particularly sized and shaped stone fell on such-and-such a day as testified by specific observers whose skill and sincerity could be granted. "In nature," Bacon wrote, "nothing really exists besides individual bodies, performing pure individual acts." He required a "collection or particular natural history of all prodigies and monstrous births of nature; of everything . . . that is in nature new, rare, and unusual." This was a programmatic justification for the "cabinets of curiosities" then fashionable in gentlemanly circles throughout Europe (fig. 20). These cabinets eloquently testified to nature's particularity and startling variety. Stuffed with rarities and oddities, such cabinets were *accessible* proof that there were indeed more things in heaven and earth than were dreamed of in traditional philosophies.

Just as seventeenth-century moderns diverged about the proper construal and philosophical role of experience, so they differed on questions of *method* to be employed in making natural philosophical knowledge. Bacon, Descartes, Hobbes, Hooke, and others expressed supreme confidence that knowledge of nature's causal structure could be secured with certainty—*if only the mind were directed and disciplined by correct method.* Method was meant to be all. Method was what made knowledge about the natural world possible and powerful, even though prescriptions for that proper method varied greatly. The mechanical metaphor that structured knowledge of the natural world also informed the means by which that knowledge could be had. Bacon wrote that the mind was not to be "left to take its own course" but must be "guided at every step; and the business done *as if*

20. *The museum of the naturalist Marchese Ferdinando Cospi in Bologna. Such museums, with their jumble of marvelous natural and artificial objects, acted as cultural magnets for local men of letters and for gentlemen making the Grand Tour of Europe. Source: Lorenzo Legati,* Museo Cospiano annesso a quello del famoso Ulisse Aldrovandi . . . *(1677).*

by machinery." For Bacon, as for many other practitioners, natural philosophy was defined by its goal of securing causal knowledge: that was what made it philosophy as opposed to history. But in this version the proper method proceeded *from* accumulated knowledge of particulars—observational and experimental facts—*to* causal knowledge and general truths; that is, it was an *inductive* and *empirically grounded* procedure. This was why the register of fact had to be secure. If the foundation was weak, the building erected on it would be shaky. Inductive procedures, and their attendant attitudes toward modes of inference from the factual register, were influential in England, though, as we shall see, many Continental practitioners, and even some English ones, expressed skepticism about their philosophical legitimacy, security, and point.

Bacon advertised his inductive method as the inversion of traditional natural philosophical practice. Hitherto, he noted, natural philosophy had tended to use particulars only as a quick means to arrive at general principles of nature. With the truth of those principles taken as indubitable, they could then be used to judge among the phenomena of nature, to decide between the often conflicting evidence of sensory experience. That method of reasoning *from* settled general principles—considered to be true in their own right—*to* the explication of particulars is called *deduction,* and to its sole use Bacon ascribed the ills of contemporary natural philosophy. It was not true, as is sometimes said, that Baconian induction commended mindless fact collecting. For one thing, Bacon and his followers made much of so-called *crucial instances,* whose purpose was to judge decisively and unambiguously between alternative physical theories.[5] For another, the senses needed to be instructed by rational method if they were to yield reliable information, fit to ground philosophy on. But in order to generate reliable causal knowledge, Bacon argued, the relative weight and priority attached to fact and theory must be radically shifted: "We must lead men to the particulars themselves; while men

5. The term "crucial experiment" appeared later in the work of Robert Boyle, referring to Pascal's dramatic Puy de Dôme experiment described in chapter 1.

on their side must force themselves for a while to lay their notions by and begin to familiarize themselves with facts." Neither sense nor reason alone could suffice to make natural philosophical knowledge, but the time had come to bring theorizing face-to-face with the facts.

And here another important qualification to the adequacy of modernist rhetoric has to be made. Despite modernist stress on the priority of direct sensory experience, Bacon also agreed with many other seventeenth-century natural philosophers that the uninstructed senses were apt to *deceive* and that the senses needed to be methodically disciplined if they were to yield the authentic factual stuff philosophical reason could work on. Just as theory uninformed by fact was to be rejected, so the disastrous state of much existing natural knowledge was often referred to the role of uninstructed sense and undisciplined sensory reports. To untutored sense, the moon looked no bigger than an apple pie and the sun appeared to go round the earth. It was educated reason, not simple sense, that allowed moderns to "see" the moon as very big and the sun as still. Few modern practitioners, however keen they might be on the foundational role of experience, therefore omitted to treat the inherent unreliability of the senses. Galileo famously applauded Copernicus for making "reason so conquer sense that, in defiance of the latter, the former became the mistress of [his] belief." Joseph Glanvill (1636–80), a vigorous publicist of English experimentalism, observed that "in many particular cases, we are not assured of the report of our Senses," needing knowledge to "correct their Informations." And Robert Hooke, himself a prolific inventor of scientific instruments, while noting the "narrowness and wandering" of the senses of fallen man, lauded the role of the telescope and the microscope in correcting their infirmities and extending their empire. (No worries here about the reliability of instrumentally mediated experience.) The progress of knowledge was referred not simply to an expanded role for sense but to a studied correction of sense by reason, perhaps by mechanical instruments, and certainly by practical procedures for assessing reports of sensory experience.

If experience was to play its foundational role in a reformed and orderly natural philosophy, therefore, it had to be controlled, monitored, and disciplined. If untutored sense was likely to mislead, then ways had to be found to regulate *what* experience could properly ground philosophical reflection. The question of *what* experience encompassed judgments about *whose* experience. The boundary between authenticated experience and what was widely called "old wives' tales" had to be marked and insisted on. The English natural philosopher John Wilkins, for example, distinguished between the "vulgar" and the learned on precisely these grounds: the former accorded primacy and privilege to immediate sense impressions, whereas the latter were suitably cautious about their reliability: "You may as soon persuade some country peasant that the moon is made of green-cheese, (as we say) as that it is bigger than his cart-wheel, since both seem equally to contradict his sight, and he has not reason enough to lead him farther than his senses." Boyle argued that the judgment of "the undiscerning multitude . . . seems rather lodged in the eye than in the brain," and that was the major basis for vulgar error. The physician Sir Thomas Browne's (1605–82) *Pseudodoxia epidemica* (1646) observed the "erroneous disposition of the people" that made them credulous and readily deceived by "fortune-tellers, jugglers, [and] geomancers." Sense needed to be guided by knowledge, and lacking knowledge the common people were "but bad discerners of verity": "Their understanding is so feeble in the discernment of falsities, and averting the errors of reason, that it submitteth unto the fallacies of sense, and is unable to rectify the error of its sensations." That is to say, for such practitioners the disciplining of experience importantly implicated a map of the social order. Experience suitable for philosophical inference had to emerge from those sorts of people fit reliably and sincerely to have it, to report it, or, if it was not their own, to evaluate others' reports of experience. Undisciplined experience was of no use.

Historians and philosophers of science have traditionally paid far too much attention to formal methodological pronouncements, often taking such statements at face value as adequate accounts of what past practitioners actually did when they went about making, assessing, and distributing scientific knowledge. In fact, the relation between *any* body of formal methodological directions and concrete natural philosophical practice in the seventeenth century is deeply problematic. For example, neither those whose methodological pronouncements professed radical disengagement between theorizing and fact gathering nor those who announced their systematic skeptical rejection of traditional culture wholly succeeded in their aims. There is much to commend a revisionist view that formal methodology is to be understood as a set of rhetorical tools for positioning practices in the culture and for specifying how those practices were to be valued. This is not, however, to deny formal methodology a role in seventeenth-century science. Methodology may be in part, as it has been called, a "myth," but myths may have real historical functions. Methodological pronouncements like Bacon's were avidly seized on by later, especially English, natural philosophers to *justify* a concerted collective program of observational and experimental fact collecting, while broadly deductive methodologies were used by other sorts of philosophers to justify the importance of rational theorizing over the accumulation of factual particulars. Formal methodology is important, therefore, in the same way that the justification of a practice is important to its recognized identity and worth. A practice without an attendant myth is likely to be weak, hard to justify, hard even to make visible as a distinct kind of activity.[6] Justifications are not to be simply equated with the practice they justify, and we still need a more vivid picture of what a range of modern natural philosophers actually *did* when they set about securing a piece of knowledge. Modern natural philosophers did not just *believe* things about the natural world; they *did* things to secure, to justify, and to distribute those beliefs. Doing natural philosophy, that is, was a kind of work. So we now need to turn from abstract methodological formulas to the practical work

6. Making a related point, sociologists might say that methodologies can be regarded as *norms*—stipulations of what conduct *ought to be*—and like all norms, they can fulfill the function of reminding people how they should behave, even if they do not describe how people always, or even usually, do behave.

of *making* experience fit for certain kinds of natural philosophical inquiry.

How to Make an Experimental Fact

Just as the mechanical metaphor lay at the heart of new strands of natural philosophy, so mechanical means came to assume a new importance in making knowledge. This stress on artificially contrived experiments is nowhere more apparent than in research programs associated with the Royal Society of London (founded in 1660) and especially with its most influential fellow, Robert Boyle. The air pump invented for Boyle by his assistant Robert Hooke in the late 1650s swiftly became emblematic of what it was to do experimental natural philosophy (fig. 21). It was the Scientific Revolution's greatest fact-making machine. How did the air pump work? How was it reckoned to make factual scientific knowledge? And how was the resulting knowledge offered as a remedy for existing intellectual ills and as an exemplar of how properly scientific knowledge ought to be produced? The following pages offer a vignette of a specific, highly influential set of knowledge-making practices, while later sections warn against the assumption that they were universally approved, even by fellow modern mechanical philosophers.

The air pump had an emblematic character in two respects: first, it and the practices mobilized around it were made into models of the right way to proceed in experimental natural philosophy. The Royal Society vigorously advertised its experimental program throughout Europe, and experimentation with the air pump was repeatedly pointed to as a paradigm of experimental philosophy. The natural philosophical use of instruments like the air pump was recognized as a new thing in the seventeenth century, attracting widespread support, imitation, and also opposition. Many histories of experimentation in natural science plausibly tell origin stories tracing back to Boyle's air pump.

Second, manipulations with instruments like the air pump could

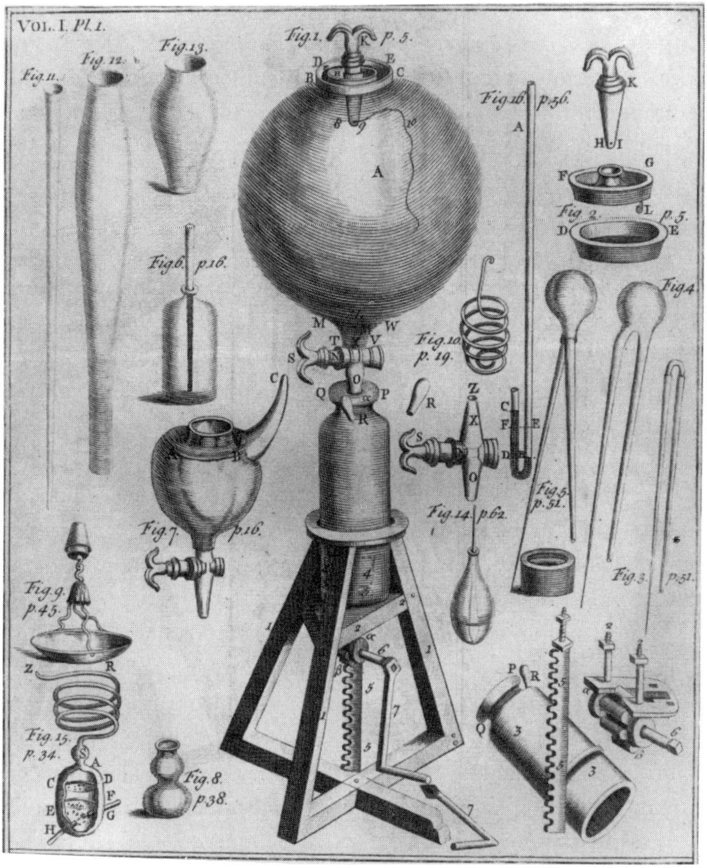

21. *Robert Boyle's first air pump. Source: Robert Boyle,* New Experiments Physico-mechanical Touching the Spring of the Air *(1660).*

yield general natural philosophical knowledge only insofar as the effects artificially produced in and by them were taken as reflecting how things were in nature. Chapter 1 discussed the general modern rejection of the Aristotelian distinction between "nature" and "art." Unless it was accepted that there was a basic similarity between the products of nature and those of human artifice, experimental manipulations with machines could not *stand for* how things were in nature, and the spread of the clock metaphor for nature, as well as the credibility of telescopic observations of the heavens, marks that acceptance. Experimentation with such instruments opened up the possibilities of enormous control and convenience. One might in principle lay on experimental phenomena at will, at any time, in front of any observers, without waiting for them to occur naturally; one might even produce effects that were not at all accessible to normal human experience. In the case of the air pump much of the natural philosophical interest in its artificially made phenomena proceeded by accepting that the vacuum it produced might stand for what would be observed if one were to travel to the top of the atmosphere. The pump might make accessible and manifest the invisible, and normally insensible, effects of the air. Yet these practical recommendations in favor of artificial experimentation were utterly dependent on accepting the principle that the products of human art could and did stand for the order of nature. Without that basic acceptance, there could be no secure inference from what experimental apparatus made manifest to the natural order of things.

The air pump was intended to produce an operational vacuum in its great glass receiver. By repeatedly drawing the piston (or "sucker") of the pump up and down and adjusting the valve and stopcock connecting the receiver to the brass pumping apparatus, quantities of air could be removed from the receiver. The effort of drawing the sucker down became more and more difficult until at last it resisted all human effort. At that point Boyle judged that he had exhausted almost all atmospheric air from the receiver. This itself counted as an experiment, and it was reported as the first of Boyle's series of forty-three *New Experiments Physico-mechanical Touching the Spring of the Air* (1660). It was this operational vacuum that was to stand for the impossible task of traveling to the top of the atmosphere, and Boyle offered a mechanical account of the tactile experience of working the sucker.

The exhausted receiver of the air pump was, however, less significant as an experiment in itself than as a space in which one might do experiments (fig. 22). The receiver had a removable brass cover at

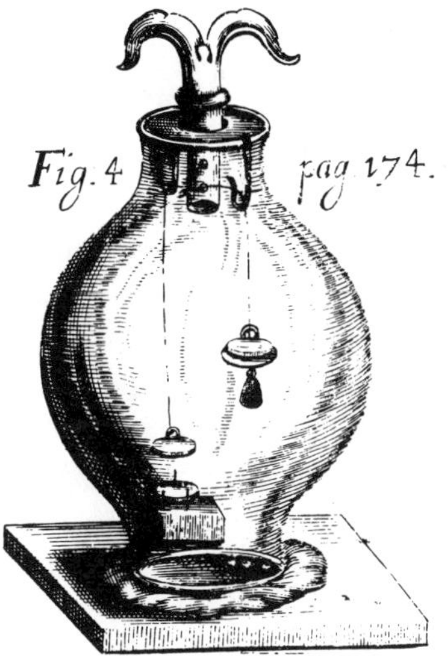

22. *An experiment in Robert Boyle's air pump. This illustration shows the receiver of Boyle's second air pump, developed about 1662. The experiment depicted involved the well-known and much-discussed phenomenon of the spontaneous cohesion of smooth marble disks. Boyle aimed to explain this effect by reference to air pressure, predicting that when the receiver was exhausted the disks would separate. Source: Robert Boyle,* Continuation of New Experiments Physico-mechanical Touching the Spring and Weight of the Air *(1669).*

the top, over an opening big enough to allow instruments to be inserted into the glass globe, and the rest of Boyle's series of trials on the properties of air consisted of observations of objects and apparatus placed in the receiver. Consider the seventeenth experiment in this series, which Boyle characterized as "the principal fruit I promised myself from our engine." This experiment consisted simply of placing the Torricellian apparatus—the mercury barometer described in chapter 1—into the receiver, which was then gradually exhausted. Boyle announced an expectation about this experiment that at once indicated its emblematic status and its role as confirming a broadly mechanical view of nature. He expected that as the receiver was exhausted the level of mercury in the barometer would fall. And when he had totally, or almost totally, exhausted the receiver, then the mercury in the long tube would descend all the way, or almost all the way, to that contained in the vessel below. If Pascal's brother-in-law had carried his barometer not just up the Puy de Dôme but to the very top of the ocean of air surrounding the earth, this is what he would have observed. And indeed, whereas there was no change in the mercury's height when the barometer was placed in the receiver and sealed up,

Boyle observed that the level of mercury descended with each "exsuction" of the pump until finally, when the receiver could be exhausted no more, it stood just a little above the level of mercury below.[7] If he turned the stopcock to let a little air back into the receiver, then the mercury reascended a bit.

Moreover, the progressive fall in the level of mercury within the receiver could not be accounted for as simply an effect of the air's *weight*—although, as Pascal and others had established, the air *did* have weight. Whereas the mercury in the vessel under Pascal's tube was exposed to the air, that in the receiver of Boyle's pump was not. It could not be said that a column of the atmosphere was weighing down on the mercury in the vessel, since the receiver's glass stood between the mercury and the atmospheric column. The weight of the body of air enclosed within the receiver itself could not be very much, certainly not enough to support a thirty-inch column of mercury. Another notion therefore seemed to be needed to account for the experimental phenomena mechanically, and this Boyle called the *pressure* or the *spring* of the air. From these and other phenomena of the pump Boyle inferred that the corpuscles of air probably have an elastic, springlike character that resists forces acting upon them and that expands when those forces are diminished. The more force you exert on an enclosed body of air, the more force it exerts back. When a quantity of air was removed from the receiver, the expansive force of the remaining air was reduced. The mercury of the enclosed barometer descended because, as Boyle said, there was then insufficient *pressure* to resist the weight of the mercury.[8]

7. Practitioners disputed—at times violently—whether the receiver was ever totally devoid of all matter or whether the pump achieved only the exhaustion of *almost all* atmospheric air. Boyle himself preferred the second account, not wishing to get embroiled in long-standing "metaphysical" debates about whether a complete vacuum could exist. He interpreted the fact that he could not get the mercury to descend all the way to mean that some residual air remained in his "exhausted" receiver.

8. Note that pressure and weight can be regarded as independent but causally related notions. In practice, Boyle was not very clear in distinguishing the two. Subsequent experimentation resulting in Boyle's celebrated "law" (inversely relating the pressure and volume of air) was undertaken in an attempt to *quantify* pressure.

The Boundaries of Natural Knowledge

The pressure or spring of the air was made visible as a major achievement of an experimental program in natural philosophy: artificial experiment was what made pressure manifest as a real force mechanically operating in nature. The artificial effects of the air pump counted as *matters of fact* about nature. Experimental facts testified to a unitary order of nature that was causally responsible for those facts. The facts of the air pump were visible or tangible, whereas the causes they testified to were not accessible to the senses. How then was it proper to move from the one to the other? How was it proper to speak respectively of matters of fact and their physical causes?

Many modern practitioners—including some (like Descartes) unimpressed with the value of a program of systematic experimentation—were agreed that the intellectual qualities of factual and theoretical knowledge differed. Here again the metaphor of the clock was drawn on to express the varying degrees of confidence one might repose in matters of fact and in hypotheses one might frame about the underlying causes of those facts. We see a clock sitting on the mantelpiece. Observing the regular movement of its hands, we have knowledge of *effects*. When reliably observed and communicated, these count as matters of fact. Such conditions being met, we can have *certain* knowledge of these facts, for practical purposes as certain as the results of mathematical or logical demonstration. But suppose the inner workings of the clock are securely enclosed in an opaque box and are practically inaccessible to our inspection. We cannot then have similarly certain knowledge of what *causes* produce these effects. That the causes are mechanical in nature is something we can be reasonably assured of (according to mechanical philosophers), but how exactly the mechanical bits and pieces are arrayed is a matter of only *probable* knowledge.[9] Our informed guesses at how the clock-

9. It was about this time that there was a notable shift in the meaning of the word "probable." Before the seventeenth century, to say that a claim was probable was to indicate that it was well attested, for instance, by Aristotle or other recognized

work produces its manifest effects have an irremediably theoretical and hypothetical character. With an actual clock, we could, if we really wanted to, take the box apart and pry into its workings. We could ask clockmakers how they do their work. But for nature we cannot do the same because we just cannot have direct sensory access to nature's hidden causal structure. We must infer the causal workings from the effects, and we cannot interrogate God, who is the great clockmaker. So in offering a probable account of how the world machine worked Descartes—who elsewhere insisted on the high degree of certainty of his mechanical explanations—said:

> Just as an industrious watch-maker may make two watches which keep time equally well and without any difference in their external appearance, yet without any similarity in the composition of their wheels, so it is certain that God works in an infinity of diverse ways, without it being possible for the mind of man to be aware of which of these means He has chosen to employ. . . . And I believe I shall have done enough if the causes that I have listed are such that the effects they may produce are similar to those we see in the world, without being informed whether there are other ways in which they are produced.

In the case of Boyle's experimental work on air this *probabilistic* attitude toward natural causes was particularly apparent in his account of the air's spring and his confidence about its cause. Boyle said he dared "speak confidently and positively of very few things, except of matters of fact," as these facts were made manifest by reliable observation or experiment. By contrast, when advancing opinions about the physical state of affairs that gave rise to those facts, Boyle recommended the utmost caution. Of such causal hypotheses he him-

---

authorities (as in our present-day sense of "probity"). By about the middle of the seventeenth century "probability" acquired a new meaning indicating an adequate degree of evidential support for a claim that was not *certainly* true.

self spoke "so doubtingly, and used so often, *perhaps, it seems, it is not impossible,* and such other expressions, as argue a diffidence of the truth of the opinions I incline to."[10] That difference in intellectual quality was even made visible in the structure of Boyle's texts. In *New Experiments* Boyle said that he meant to leave "a conspicuous interval" between his factual narratives of what the air pump had made manifest and his occasional "discourses" on their causal interpretation. Readers were invited, if they wished, to read the experiments and the interpretative "reflexions" separately.

Boyle assured readers that he had not set out to perform his experiments with a view to proving or disproving any system of philosophical grand theory. So far from being a champion of any particular body of causal theory, he said he had scarcely even read the philosophical works of such eminent natural philosophic systematists as Descartes, "that I might not be prepossessed with any theory or principles." He professed himself "content to be thought to have scarce looked upon any other book than that of nature." This was Boyle's way of signaling that theoretically interested observation was in danger of being distorted and unreliable. And the "systematical" way of proceeding—engaging with factual evidence as it bore upon an entire system of natural philosophy—was identified as a cause of the failure of traditional philosophical practice.[11]

Accordingly, Boyle said that his "business" in the air pump experiments was "not to assign the adequate cause of the spring of the air, but only to manifest, that the air hath a spring, and to relate some

10. Here Boyle was markedly more cautious than his professed methodological model, Francis Bacon, who reckoned that *certain* knowledge of physical causes *was* possible and was the legitimate goal of natural philosophy.

11. The *rhetorical* character of such stipulations needs to be stressed: it identified the proper source of authority for scientific claims. The evidence is that Boyle was, in fact, quite widely read in the systematic natural philosophical literature. In this connection he was commending a loosening of traditionally strong ties between observation and formal theorizing. It is unlikely that any absolute break between the two is possible, and Boyle *must* have approached his experimental work with a set of theoretical expectations or else he would not have been able to distinguish experimental failure from success, still less recruit his observations in support of what he called the corpuscular or mechanical hypothesis.

of its effects." To be sure, Boyle did offer some hypotheses about the corpuscular realities that were the cause of the air's spring, but these opinions were flagged with the appropriate cautions. The corpuscles of air *might* have a structure like ordinary coiled metallic springs, or they might be like sheep's fleece or sponges, or again, spring might be accounted for by the vortices Descartes favored. Such causal conjectures were surely a proper part of experimental natural philosophy, but they were less certain than matters of fact, and they were obliged to follow from the establishment of an adequate body of factual knowledge.[12] In general practice this meant that Boyle professed himself to be a mechanical philosopher but never actually offered specific mechanical explanations of physical phenomena. As chapter 1 noted, although he expressed overall confidence in explanations that were micromechanical in form, Boyle and his followers differed notably from Descartes in declining to spell out the precise micromechanical sizes, shapes, arrangements, and states of motion that gave rise to such observed qualities as spring, color, smell, and so forth. This was an in-principle mechanism, bounded by the relative grades of certainty it was right to have in factual versus causal-theoretical knowledge.

Although Boyle acknowledged that the search for causal understanding—however conditional—was an appropriate task for experimental natural philosophy, there were other bodies of knowledge that had to be excluded altogether from the business of experimental natural philosophy. If matters of fact were to count as the secure foundations of a reformed natural philosophy, they had to be guaranteed as authentic and protected from contamination by other less certain and less incontrovertible items of knowledge. The general—although not universal—tendency of seventeenth-century

12. Boyle himself did not always observe the boundaries between fact and theory that he advertised. At some points he treated the air's spring as a causal explanation of experimental effects, while at others he offered spring as a matter of fact securely evinced by experiment. Nor did Boyle ever attempt to write down the rules for how one was to move, however provisionally, from matters of fact to their mechanical explanation.

English practice was to reject the legitimacy within natural philosophy of explicitly theological, moral, and political considerations. The Book of Nature that the modern natural philosopher read was understood as God's book, but it was often said that a mechanical philosophy had to treat nature in its mechanical aspects. So, for example, in the 1660s, critics of Boyle's work on the spring of the air challenged the adequacy of a mechanical account and urged the necessity of taking spiritual powers into consideration. Boyle responded by noting his own profound piety but reminding his critics of the *proper boundaries* of natural philosophy:

> None is more willing [than myself] to acknowledge and venerate Divine Omnipotence, [but] our controversy is not what God can do, but about what can be done by natural agents, not elevated above the sphere of nature, . . . and in the judgment of true philosophers, I suppose [the mechanical] hypothesis would need no other advantage . . . than that in ours things are explicated by the ordinary course of nature, whereas in the other recourse must be had to miracles.

The power of God and spiritual agencies in the natural order was freely acknowledged, but in the view of Boyle and his colleagues the scope of *natural* philosophy was to be circumscribed by the mechanical means God had used to create the world-clock and the mechanical manner of its functioning.[13] A factually grounded experimental natural philosophy held out the prospect of a well-founded certainty and a well-conceived approach to knowledge of nature's underlying causal structure. It was widely considered that theological, moral, metaphysical, and political discussions had generated divisiveness and conflict. If a reformed natural philosophy was to offer a genuine certainty, then the demarcations between it and

13. Chapter 3 will both amplify and qualify this conception of natural philosophy by noting important theological uses of a mechanically construed nature and the extent to which pure mechanism was considered to describe adequately a range of natural phenomena.

contentious areas of culture had to be made clear. The "progress of human knowledge," Boyle wrote, had been hindered by introducing "morals and politicks into the explications of corporeal nature, where all things are indeed transacted according to laws mechanical." Put another way, the conditions of such intelligible and objective knowledge about nature as it was possible for people to secure were the separation of natural philosophy from forms of culture in which human passions and interests were engaged and the construal of nature in its mechanical aspects. To speak intelligibly and philosophically of what is "natural" or "corporeal" *was* to speak in mechanical terms. This is not necessarily the same thing as saying that mechanism was wholly adequate to account for all phenomena that present themselves to human experience. Where modern practitioners importantly disagreed was in their identification of what phenomena *were* natural.

Making Knowledge Public

It is traditional to trace the contours of the Scientific Revolution through the texts of individual practitioners. Yet the individual natural philosopher did not make knowledge all alone, and the very idea of *knowledge* implicates a public and shared commodity, to be contrasted with the individual's state of *belief.* To establish its credibility and to secure its status as knowledge, individual belief or experience has to be effectively communicated to others. Indeed, modern natural philosophers devoted much reflective attention and practical work to the question of just *how* experience could effectively and reliably make the passage from the private to the public domain. Many practitioners judged that the widely diagnosed sickness of contemporary natural philosophy proceeded from its excessively private or individualistic character, and the next chapter will consider some dangers that were seen to flow from intellectual individualism and privacy.

We have seen that the seventeenth-century English empirical tradition laid special emphasis on factual particulars as the secure foundation for natural philosophical knowledge. If particular experiences were to serve that function, however, their authenticity as actually occurring, historically specific happenings somehow had to be guaranteed and made persuasive to a community. Consequently, if such particulars were to become part of a shared stock of knowledge, reliable ways had to be found to make them travel, to extend them from an individual to many others. Boyle and his associates developed a variety of relatively novel techniques to assist the transition of experimental and observational experience from the individual to the public domain. First, recall that one of the recommendations of a program of experimentation was greater practical control over experience. Unlike purely natural phenomena, those produced in instruments like the air pump could—within practical limits—be laid on when and where one liked. Witnesses could be specially mobilized to observe experimental effects and to attest their authenticity, and Boyle's texts sometimes named witnesses to his air pump trials. Moreover, experimental performances were a routine feature of the meetings of the Royal Society, and a *Register-Book* was provided for witnesses to testify their assent to experimental results. Second, Boyle influentially recommended that experimental reports be written in a way that allowed distant readers—not present as firsthand witnesses—to *replicate* the relevant effects. Actual methods, materials, and circumstances were to be minutely detailed so that readers who were of a mind to do so could reproduce the same experiments and thus become direct witnesses.

As it happened, neither of these techniques was a very powerful means of extending experience. For practical reasons alone the number of direct witnesses for experimental performances was always limited: in Boyle's laboratory that public probably consisted of at most three to six competent colleagues, and audiences for Royal Society trials rarely exceeded twenty and were typically much smaller. And though Boyle's texts encouraged replication and offered detailed instructions on how to proceed, after a number of years even he became convinced that few competent replications of his air pump experiments had been carried out and concluded that few were ever likely to be. Accordingly, if experience was to be effectively extended, means other than public witnessing and physical replication had to do the work.

Such means were found in the forms of scientific communication itself. Experience might be extended and made public by *writing* scientific narratives in a way that offered distant readers who had not directly witnessed the phenomena—and probably never would—such a vivid account of experimental performances that they might be made into *virtual witnesses.* Most practitioners who took Boyle's factual particulars into their stock of knowledge did so not through direct witnessing or through replication but through reading his reports and finding adequate grounds to trust their accuracy and veracity. As Boyle said, his narratives (and those that competently followed the style he recommended) were to be "standing records" of the new practice, and readers "need not reiterate themselves an experiment to have as distinct an idea of it, as may suffice them to ground their reflexions and speculations upon." Virtual witnessing involved producing in a reader's mind such an image of an experimental scene as obviated the necessity for either its direct witness or its replication. In Boyle's experimental writing this meant a highly *circumstantial* style, often specifying in excruciating detail when, how, and where experiments were done; who was present; how many times they were reiterated; and with exactly what results. Experiments were to be detailed in great numbers, and failures were to be reported as well as successes. Such a prolix style might "keep the reader from distrusting" the outcomes related and might assure the reader of the specific historical reality of factual particulars.

The scientific author appeared as disinterested and modest, not concerned for fame and not affiliated with any school of grand philosophical theorizing: "It is none of my design," Boyle wrote, "to engage myself with, or against, any one sect of Naturalists." Such a person could be believed, and such a person's narratives might be treated as the transparent testimony of nature itself. A circumstantial and a sincere way of writing might transform readers into witnesses. Experience could be extended and the factual grounds of natural philosophical practice could be made more secure. Once the factual foundations of natural knowledge had been guaranteed by these means, the philosophical search for causes could safely proceed.

What Is the Point of Experiment?

The accumulation of particulars as a way of grounding a reformed natural philosophy was one important strand of modern practice, particularly favored by the English but influential on the Continent as well. Publicized through a network of scientific "intelligencing" centered on the Royal Society of London, broadly experimental and inductive practices secured a foothold in a number of European countries and even in the emerging scientific culture of the American colonies. Yet this way of securing natural knowledge was by no means unopposed, and some moderns rejected it in whole or in part. Neither the systematic performance of experiments nor the view that a mass of authenticated factual particulars provided the uniquely secure foundations of natural philosophical knowledge simply defined what it was to be modern.

Descartes, for example, reckoned that the grounds of proper natural philosophical knowledge were to be sought through rationally conducted skepticism and self-interrogation. Doubt all you can, and when you arrive at principles that you cannot doubt then you have the foundations of philosophy. Descartes did not himself perform a great mass of physical experiments, and though he formally expressed the wish that "an infinitude of experiments" be made, he did not consider that formulating a secure natural philosophy had to await their outcomes. Experiments were to have a role, but it was not necessary to pile them up in a great heap of particulars, still less to expect to induce secure general physical principles from that heap. Descartes even complained that those experiments that had been recently communicated contained *too much* historical specificity and particularity to be philosophically useful: "They are for the most part so complicated with unneeded details and superfluous ingredients that it would be very difficult for the investigator to discover their core of truth." Unlike Boyle and his colleagues, Descartes was skeptical that a community could ever find the moral and literary means to ensure the reliability of a mass of experimental reports.

In England Thomas Hobbes took violent exception to the experimental program associated with Boyle and the Royal Society. In his view it was a pointless program. Why perform systematic series of experiments when, if one could in fact discern causes from effects, a single experiment should suffice? Nor was it evident to Hobbes that artificial effects, such as were produced by Boyle's air pump, were either necessary to natural philosophy or worth the expense and bother of performing the experiments: "What I want of experiments you may supply out of your own store, or such natural history as you know to be true; though I can be well content with the knowledge of causes of those things which everybody sees commonly produced."[14] Nor could any intellectual enterprise entitled to the name of philosophy be content with the causal caution associated with the experimental program. Referring to the Royal Society, Hobbes wrote, "They can get engines made [and] recipients made, and try conclusions; but they are never the more philosophers for all this." Through the 1660s and 1670s Hobbes sought to show the insecurity of Boyle's program of systematic experimentation with the air pump, offering detailed alternative accounts of the pump's effects and of theoretical inferences from them.[15]

A program of systematic fact collection could form a register of natural and artificial effects: it could count as natural *history*. But Hobbes insisted on the traditional construal of natural *philosophy* as the quest for secure knowledge of nature's causes—"philosophy [is]

14. In this way Hobbes, who was (as was briefly noted earlier in this chapter) one of the seventeenth century's most vehement critics of Aristotelianism, nevertheless evidently shared Scholastic attitudes about the value of commonly available experience and the problems attending esoteric experience.

15. Although the early Royal Society circle was content with the reliability of matters of fact produced and attested by Boylean procedures, such satisfaction was not universal, and a number of important Continental philosophers, as well as Hobbes, expressed doubt that what counted as a fact for Boyle and his allies was indeed a fact.

the science of causes"—and he saw no secure way of proceeding from a pile of particulars to causal knowledge that possessed the certainty appropriate to philosophy. To be entitled to the name "philosophy," a practice just could not affect Boylean caution concerning the causal structure of nature. Instead it had to proceed *from* rationally established correct knowledge of causes *to* knowledge of effects. So Hobbes refused to accept the legitimacy of Boyle's diffident attitude to the cause of the air's spring. State on certain grounds what the real cause is, and then you will be acting like a philosopher. Decline to do so and you will be a mere teller of stories about natural phenomena. The evaluation of the experimental program therefore flowed from a view about what the proper products of natural philosophical knowledge ought to be. What should be the relative knowledge-making roles of experience and rational thought? How much certainty should one expect of inquiries into the natural order? Where—between factual particulars and theoretical items—was certainty to be located? And what kind of certainty ought one to expect from genuine philosophical inquiry?

Although Hobbes was a mechanical philosopher, and though he lived and worked in England most of his life, his dismissal of the experimental way and his pungently polemical style meant that he never became a fellow of the Royal Society. Through the 1670s, a broadly Boylean experimental and natural historical program continued to characterize the Royal Society's collective work and cultural image. And though, as we have seen, Boylean corpusculariaism was in principle compatible with a mathematical approach to natural inquiry, in fact Boyle himself expressed serious reservations about mathematical idealizations, and his own experimental work was notably free of mathematical schemes and representations. This includes the "law" relating the pressure and volume of air for which Boyle is best known to modern science, a law that Boyle never called a law and to which he never gave symbolic mathematical expression.

Many accounts of the Scientific Revolution represent Sir Isaac Newton as bringing to maturity the mechanical and experimental program associated with his older Royal Society colleague Robert Boyle. Indeed, much effort was expended in England to display continuity between the Boylean program, dominant in the 1660s and 1670s, and the Newtonian program that became increasingly influential in later decades. Nevertheless, there were important divergences in how Boyle and Newton went about securing natural knowledge, in their views about the quality of certainty that was to be expected of the results of physical inquiry, and in their conceptions of the proper role of experience in natural philosophy. The experimental program of the early Royal Society was dedicated to the reform of natural philosophy through curing *dogmatism.* When practitioners had been well instructed what degree of confidence they ought to have in various types of knowledge, then natural philosophy would be securely founded and set on the right road to progress. Leading members of the Society endorsed Boyle's view that practitioners ought to repose great confidence in well-attested matters of fact while adopting a more circumspect attitude toward causal claims. Fact-founded causal knowledge was endemically incapable of the kind of certainty associated with mathematical demonstration, and those who expected physical inquiry to yield causal certainty on the model of pure mathematics were labeled deluded dogmatists. They stood accused of a category mistake—conflating inquiries into real sensible matter and its effects with the abstract realm of mathematics. The sooner natural philosophers appreciated the provisional and probable character of their theoretical accounts the better.

It was against this background that some of Newton's first contributions appeared to several notable Royal Society practitioners not as the fulfillment of the same natural philosophical program as they were engaged in but as a reappearance of discredited dogmatism. The experimental series in question was referred to at the time as the *experimentum crucis* (crucial experiment), for it purported to discriminate decisively between rival theories of the nature of light. Optics—the study of the properties and behavior of light—was less readily assimilated to a mechanical framework than, for example, the aero- and hydrostatic phenomena investigated by Pascal and Boyle. Nevertheless, in the seventeenth century much effort was expended to develop a mechanical theory of light.

The spectrum of colors—like the rainbow—that was produced when sunlight was refracted through a prism was well known.[16] Before the seventeenth century it had been customary to treat color and light as separate topics. Different-colored bodies had traditionally been thought of as having distinct real qualities—redness, yellowness, and so on—and the ensuing tension with the distinction between primary and secondary qualities spurred mechanical philosophers to develop some theory that obviated the necessity of attributing different real qualities to different-colored bodies, and thus to merge the understanding of color and light. In the 1630s Descartes made a major attempt at a mechanical theory of light, treating light as a pressure effect in a universe full of fine spherical bits of matter and color sensations as caused by the spheres' different speeds of axial rotation. Refraction through a prism, in Descartes's scheme, modified the rotation of the matter making up pure white light, and the extent of modification caused the color patterns we observe. So, while Descartes offered a mechanical account of light and color, he preserved a traditional commonsense assumption that the primitive state of light was "white" (that is, natural light) and that colors—such as those produced by refraction through a prism—were to be accounted for as modifications of "whiteness."

Newton's crucial experiment consisted in arranging *two* prisms in such a way that only one of the colored rays produced by the first refraction was refracted a second time (fig. 23). If traditional theory about the primitive nature of white light was correct, then a second refraction should cause a color change. If, however, as Newton suggested, white light was itself a mixture of different-colored rays, then the color of the ray subject to a second refraction should remain the same, and this is what Newton found to be the case. Each type of ray was concluded to have a specific refrangibility. Although contemporaries confronted immense problems in settling the identity of the cru-

16. "Refraction" designates the bending of light as it passes from one transparent medium to another, say from air to glass. "Refrangibility" refers to the capacity of different forms of light to be bent or the different capacities of media to bend light.

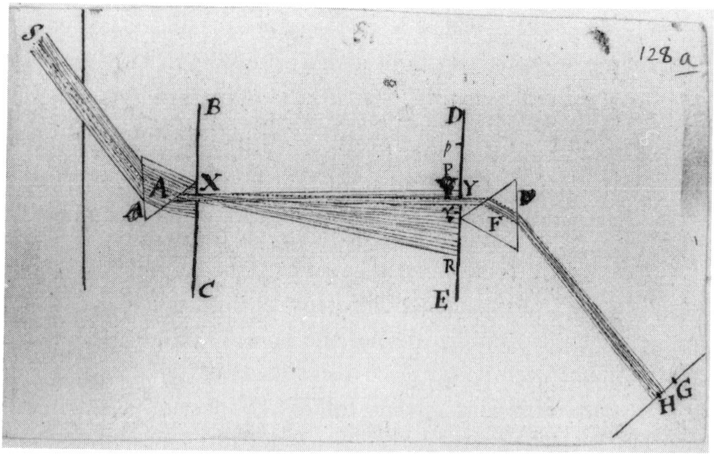

23. *Isaac Newton's "crucial experiment." This drawing, from the manuscript of Newton's opitcal lectures as Lucasian Professor of Mathematics at Cambridge, shows an early version of the two-prism experiment. Source: Cambridge University Library MSS Add. 4002, fol. 128a.*

cial experiment, the most substantial trouble involved Newton's claims about *what* this experiment established, *how* it established a theory of light, and with what *certainty* it did so.

Unlike Boyle's experimental reports, Newton's communications to the Royal Society in the early 1670s about his work with prisms offered only the most sketchy accounts of manipulations and their circumstances. Although the experiments concerned were presented as decisive, their reporting was far from detailed. Indeed, Newton acknowledged the relatively stylized manner of his experimental reporting, saying in mitigation that "the historical narration of these experiments would make a discourse too tedious & confused, & therefore I shall rather lay down the *Doctrine* first and then, for its examination, give you an instance or two of the *Experiments,* as a specimen of the rest." Later he justified the sparseness of his experimental narratives by drawing an implicit contrast with Boylean practice: "It is not number of Experiments, but weight to be regarded; & where one will do, what need of many?" Clearly, historically particular experience was not meant to play the role in Newton's natural philosophy that it did in Boyle's.

Moreover, in Newton's initial communication of these results to the Royal Society in 1672 he claimed to have ascertained "the true cause" of optical phenomena, and to have done so with *certainty:*

> A naturalist would scarce expect to see the science of [colors] become mathematical, and yet I dare affirm that there is as much certainty in it as in any other part of optics. For what I shall tell concerning them is not an hypothesis but most rigid consequence, not conjectured by barely inferring 'tis thus because not otherwise, or because it satisfies all phenomena . . . but evinced by the mediation of experiments concluding directly and without any suspicion of doubt.

In this connection "the true cause" of a prismatic image to which Newton referred was the interpretation of light as consisting of differently refrangible rays. Yet that cause was also there associated with a corpuscular theory of the physical nature of light rays, a theory that Newton elaborated through the 1670s and that was in keeping with his overall commitment to a mechanical metaphysics.

If Newton was taken to assert that he had established a physical causal claim with "certainty" and "without any suspicion of doubt," then this was exactly the kind of thing that Royal Society experimentalists had learned to reject as dogmatism. Boyle's associate Robert Hooke upbraided Newton on just these grounds.[17] Granting the reliability and veracity of Newton's experimental narratives, and granting that Newton's hypothesis *could* explain his findings, how

17. But *was* Newton actually offering such a claim? Under pressure from Hooke, Newton denied that he was. Here he said that he was setting aside questions of underlying causal mechanisms or that he was advancing such mechanisms only hypothetically, as in the case of his views on gravitation noted in chapter 1. He said that he declined to "mingle conjectures with certainties." Yet Hooke's presumption was not unfounded: Newton's notebooks of the period show an intense commitment to corpuscular theories of the physical nature of light about which his public posture was inconsistent.

could it be said that physical investigations could *prove* the truth of any theory that purported to account causally for factual matters? If, as the clock metaphor suggested, we infer from manifest facts to hidden causal structure, then we have to accept that a *number of causal theories* can explain the same facts. There is no proof but only probability in such inferences, and Hooke said he had an alternative optical theory that could account for the same effects "without any manner of difficulty or straining." Newton's was, Hooke confessed, an "ingenious" hypothesis, "but I cannot think it to be the only hypothesis; nor so certain as mathematical demonstrations." Newton stood accused of offending against the modesty and good manners appropriate to proper natural philosophy in Boyle's idiom.

But Newton, it might be said, had not so much violated the rules of one game as he had insisted on the legitimacy of playing by the rules of another game. The certainty of mathematical demonstration was what he was after, insofar as it could be legitimately attained in physical inquiry. He was not content with probability, and he did not accept Boylean limits on the certainty to be had in natural philosophy. He hoped that "instead of the conjectures and probabilities that are being blazoned about everywhere, we shall finally achieve a natural science supported by the greatest evidence." Newton's expectations of physical certainty arose from the mathematical rather than experiential foundations of his natural philosophical practice. He rejected physical theories unless they could be mathematically "deduced" from experiment, but those theories that could legitimately be so deduced were to be spoken of with absolute confidence, not with the caution of the probabilist.[18] The aim—so far as it was possible—was to bind assent in iron chains of mathematical and logical deduction,

18. The term "deduction" was Newton's own, yet its appropriateness was hotly contested by a number of contemporaries. Deduction was widely taken to imply that there was no room for negotiation or disagreement about inference from experiment, whereas critics who accounted themselves quite rational saw adequate grounds for such inferential dissent. Indeed, Newton's prism experiments were not at all easy for others to replicate, and some practitioners who tried and failed treated the "matter of fact" with skepticism.

seeking to guide the mind along from necessary truth to necessary consequence.

The confrontation over Newton's optical work can stand as an emblem of the fragmented knowledge-making legacies of the seventeenth century. A theoretically cautious and experience-based conception of science was here juxtaposed to one that deployed mathematical as well as experimental tools to claim theoretical certainty. Diffidence was opposed to ambition, respect for the concrete particularities of nature to the quest for universally applicable idealizations, the modesty of the fact gatherer to the pride of the abstracted philosopher. Do you want to capture the essence of nature and command assent to representations of its regularities? Do you want to subject yourself to the discipline of describing, and perhaps generalizing about, the behavior of medium-sized objects actually existing in the world?

Both conceptions of science persist in the late twentieth century, and both can trace elements of their formation back to the seventeenth century. The one is not necessarily to be regarded as a failed version of the other, however much partisans may defend the virtues of their preferred practice and condemn the vices of another. These are, so to speak, different games that natural philosophers might wish to play, and decisions about which game is best are different in kind from decisions about what is a sensible move within a given game: an accurate pass from midfield to the winger in soccer is not a bad jump shot in basketball. In the seventeenth century natural philosophers were confronted with differing repertoires of practical and conceptual skills for achieving various philosophical goals and with choices about which ends they might work to achieve. The goal was always some conception of proper philosophical knowledge about the natural world, though descriptions of what that knowledge looked like and how it was to be secured varied greatly.

Three

WHAT WAS THE KNOWLEDGE FOR?

Natural Philosophy Cures Itself

Seventeenth-century mechanical philosophers attempted to discipline, if not in all cases to eliminate, teleological accounts of the natural world. Yet as ordinary actors they accepted the propriety of a teleological framework for interpreting human cultural action, and with some exceptions so do modern historians and social scientists: the very identity of human action—as action rather than behavior—embodies some notion of its point, purpose, or intention. An account of the action of someone waving good-bye is not adequately given by detailing the muscular movements involved. Similarly, any interpretation of what natural philosophers believed and what they did has to deal with the *purposes* of natural knowledge. In general, what was natural knowledge *for?* Specifically, to what ends was a *reform* of natural knowledge undertaken in the seventeenth century? Natural knowledge was given its shape in contexts of purposive use, and its meanings emerged from its uses.

One may take it as a matter of course that early modern natural philosophers as a group were in part motivated by a desire to produce and extend true, or probably true, knowledge. Arguably, so are all scholars worthy of the name, of all types and at all times. Just because that motive is plausibly general, it cannot effectively *discriminate* between strands of practice—ancient versus modern, mechanical versus animistic, inductive versus deductive, and so on. Therefore if one wants to treat *changes* in natural philosophy, or to account for different *versions* of natural philosophy, the motive of "the search for truth" is of no use at all. One needs to look for purposes that distinguished one type of practice from another and for the varying predicaments in which those purposes were acquitted. Moreover, whatever motives may have moved individual practitioners cannot be sufficient to account for either the credibility of their knowledge or the legitimacy it secured in their societies. Thus there is every reason to accept "the desire for knowledge" as among the motives of early modern natural philosophers, while setting that consideration to one side in interpreting *changes* and *differentiation* in natural philosophical knowledge, as well as in understanding the bases for social acceptance and approval of their knowledge. Among the things that varied were conceptions of what genuine knowledge of nature *is*.

During the course of the disputes occasioned in the 1670s by his "crucial experiment" on the nature of light, Newton expressed exasperation at his inability to secure from other practitioners the sort of unconditional assent he hoped and expected that his new mathematical natural philosophy would achieve. Despairing, he threatened to give up philosophy altogether. In fact it was natural *philosophy*—the inquiry aiming at knowledge of physical causes—that he said he would abandon, not mathematics—the study of the regularities that might be discerned in natural phenomena, whatever their physical causes. Mathematics evidently could deliver on its promises of demonstrative certainty, whereas natural philosophy apparently remained—despite Newton's best efforts to mathematize its practice—notably incapable of securing universal agreement and allaying doubt. Newton considered that natural philosophy *ought* to offer a high degree of certainty and that its formal procedures *ought* to ensure universal assent. But as it was then constituted, it did not. Philosophy, Newton complained, was an "impertinently litigious Lady."

Newton was here echoing a widespread sixteenth- and seventeenth-century modern sentiment about the cultural characteristics of natural philosophy. In its traditional forms, especially, it was seen as a notoriously divisive and disputatious form of culture. Descartes wrote that philosophy had been "cultivated for many centuries by the best minds that have ever lived," but nevertheless "no single thing is to be found in it which is not [the] subject of dispute." One English contemporary commentator described academic dissent and disputatiousness in the early seventeenth century as "an Epidemick evill of that time." Moderns saw traditional systematists incessantly warring with each other, ignorant armies clashing by night, producing nothing solid or constructive and effectively achieving neither conversion nor consensus. Critics portrayed traditional natural philosophy as a scandalous school, its disorderliness to be taken as a reliable token of its corrupt state. No house so divided against itself could or should long stand, and chapter 2 noted modern judgments that the whole edifice should be torn down and begun anew. Accordingly, modern methodological and practical reforms aimed overwhelmingly at curing natural philosophy of its existing ailments and specifically at rectifying its infamous disorder. So the first purpose that might be served by a reform of natural philosophy was the cure of its own body. Permitted to go untreated, a sick natural philosophy could secure no credibility in the culture and fulfill no other social or cultural purposes.

The disputatiousness of traditional natural philosophy was frequently blamed on the dominant role of university scholars and traditional scholarly ways of establishing and justifying knowledge. A typical form of philosophical exchange in the universities was the ritual disputation, in which opposing scholars deployed sophisticated logical and rhetorical tools to defend their theses and defeat those of their opponents, the results to be judged by a philosophical master. So when moderns insisted they would mind not words but things, they were referring quite specifically to the verbose and wrangling style of natural philosophy in the Schools. The litigious and disputatious manner of traditional scholars was ridiculed by civil society in general as well as being scorned by modern natural philosophers. And if philosophical reformers diagnosed traditional wordiness and litigiousness as symptoms of intellectual disease, mercantile as well as polite society often saw the wrangling scholar as a figure of fun, useless to civil society. An English critic of the universities in the 1650s indicted Scholastic culture as "a civil war of words, a verbal contest, a combat of cunning craftiness, violence and altercation, wherein all verbal force, by impudence, insolence, opposition, contradiction, derision, diversion, trifling, jeering, humming, hissing, brawling, quarreling, scolding, scandalizing, and the like, are equally allowed, and accounted just."

Chapter 2 introduced several of the cures that moderns proposed for natural philosophical divisiveness and disorder. One was systematically formulated *method,* seeking by explicit rules of reasoning and the control of experience to ensure that all participants were, so to speak, on the same page in the Book of Nature. It was hoped that imperfections in sensory capacity, variations in "wit," and divergences in theoretical or social interest might all be corrected by the mechanical action of *right method.* When all proceeded in the same way, and when all accepted the same stock of knowledge, then philosophical disorder would be truly healed. But the scholarly way of life had been disputatious and wrangling for centuries. Why was this mark of disorder recognized in the late sixteenth and seventeenth centuries as particularly in need of remedy? The answer requires some attention to changing boundaries of participation in natural knowledge and changing patterns of interest in it.

In the middle of the sixteenth century the preface to Copernicus's *De revolutionibus* carefully circumscribed the audience for his text: "Mathematics," he said, "is written for mathematicians." In 1600 William Gilbert announced his cavalier disregard for common opinion: "We care naught, for that, as we have held that [natural] philosophy is for the few." Galileo vigorously endorsed those exclusionary sentiments, attempting to drive a wedge between the perceptions and competences of "the common people" and those appropriate to genuine mathematical and natural philosophical expertise. The delivery of truth about the natural world was to be the preferential preserve of those who possessed special competence. Despite much early Royal Society rhetoric stressing the virtues of a more open natural philosophical practice, social realities remained substantially restrictive. Artisans, for example, were rarely represented in the new scientific societies springing up throughout Europe—if not always for reasons of overt social distaste, then because criteria of intellectual competence presupposed a course of formal education through which the masses seldom passed. The view that mathematics, and much of natural philosophy, ought to be produced by and for certifiably expert practitioners continued to be an important sentiment right through our period of interest and beyond. There was nothing new in this. Historians appreciate that even in antiquity there was a gulf of competence and comprehension separating the mathematical sciences—including much that bore on interpreting the physical world—from the understanding of even ordinarily educated people. The book that is said to have marked the culmination of the Scientific Revolution and to have changed the way "we" think about the world—Isaac Newton's *Principia mathematica*—was probably read in its entirety by fewer than a hundred contemporaries, of whom no more than a handful were competent to understand it.

Natural Knowledge and State Power

The most diffuse and difficult to summarize, but probably the most far-reaching, links between natural knowledge and state power flowed from broad European changes in attitudes to knowledge in general and to the relations between knowledge and social order. The environment for these changes was what might be called a state of *permanent crisis* affecting European politics, society, and culture from the late medieval period through the seventeenth century. Some markers of that continuing crisis include the breakdown of the feudal order and attendant rise of strong nation-states from the thirteenth century onward; the discovery of the New World and both the cultural and the economic shocks emanating from that expansion of horizons; the invention of printing and consequent changes in the boundaries of cultural participation; and the fragmentation of a unified Western European religious order that followed from the Protestant Reformation of the sixteenth century. Each of these events, but especially the last, eroded the authority and the effective scope of institutions that had regulated human conduct for preceding centuries. The Roman Catholic papal authority that had—formally at least—unified Western Europe under a single Christian conception of authority gave way to split sources of authority: clashes first between divine and secular notions of political authority, then between different versions of Christianity and their proper relation to secular political authority. The wars of religion between Catholics and Protestants that raged across Europe from the Reformation onward, but particularly the Thirty Years' War of 1618–48, were the immediate occasion for changed views of knowledge and its role in ensuring or subverting order.

When systems of institutional control are working without significant challenge, the authority of the knowledge embodied in the institutions seems similarly potent. When the institutions are attacked and then fragment, however, problems about knowledge and its legitimacy come to the fore. In such circumstances, *skepticism* about current systems of knowledge may flourish, for little about existing intellectual systems seems self-evidently satisfactory any more. What is proper knowledge? What guarantees its truth? What degree of certainty can we have, and is it proper to expect, of knowledge? Who can have knowledge and on what conditions? Can people be made to believe the same things and, if so, by what purposefully designed means? Since social order is seen to depend on shared belief, what criteria of right thinking can be displayed and implemented to ensure such consensus? Candidate solutions to these questions are proposed and their merits debated.

It is just when the authority of long-established institutions erodes that the solutions to such questions about knowledge come to have special point and urgency. It is concluded that existing techniques for securing knowledge are evidently inadequate, and new procedures are canvassed. Method, broadly construed, is the preferred remedy for problems of intellectual disorder, but which method is it to be? Here the overarching problem—to which correct method is supposed to be an answer—is skepticism as the solvent of all secure belief. How to bound skepticism? How to manage it within safe limits? How even—with Descartes—can one turn skepticism on itself and, by taking it to its limits, show that what remains is immune to doubt? Debates over method take on greater significance when it is reckoned that the order of society depends in large measure on arriving at and then disseminating the correct method for securing belief. Addressing the problem of knowledge, and the skepticism that corrodes belief, is what links philosophers' work to the concerns of the wider society. This permanent crisis of European institutions during the early modern period affected attitudes toward knowledge in general, and it affected attitudes to natural knowledge for reasons that were touched on in earlier chapters and will be developed below. Knowledge of nature was considered deeply relevant to problems of order, not least because nature was widely understood to be a divinely written book whose proper reading and proper interpretation had the potential to secure right belief and thus to guarantee right conduct. Conversely, right belief and conduct could always potentially be subverted by improper ways of reading and interpreting the Book of Nature.

The permanent crisis of European order was, then, the general backdrop to debates over natural knowledge and its relation to state power and social order. Yet appreciations of that relationship were also shaped by more specific European developments, one set of which concerned changes in the sorts of people who participated in natural knowledge and associated changes in expectations about what natural knowledge was good for. If natural philosophy remained the exclusive concern of professional scholars, there would be no particular reason to suggest that its contentiousness required urgent remedy. Medieval and early modern scholarly life just *was* contentious, and few university scholars saw anything very wrong in this. Yet interest in natural knowledge was *never* the sole prerogative of university scholars, and in the sixteenth and seventeenth centuries relatively new social and cultural considerations began to strongly affect the practice of natural philosophy and natural history.

From the medieval period right through the seventeenth century many, and probably most, natural philosophers were clerics or worked in institutions controlled by or allied with the church, such as the universities themselves. Some of those not formally affiliated with religious institutions enjoyed church patronage, while few could, or wished to, separate their scientific work from church concerns. But throughout the early modern period other sources of support for, and interest in, natural knowledge were developing, relatively disengaged from formally religious preoccupations.

One current was associated with princely courts, especially on the Continent. Here sovereigns offered mathematicians, astronomers, natural historians, and natural philosophers patronage that proved vital for the structure of a number of important scientific careers. Recent historical work on Galileo, for example, has stressed the significance of court patronage relationships not only for his livelihood but also for the thematics and presentation of his scientific work. "Court philosophers" might shed cultural luster on highly competitive and prestige-conscious Italian princes: Galileo knew very well how much it was worth to his Florentine patrons, and to his own career, to name his newly discovered moons of Jupiter the "Medicean stars" for the ruling Medici family. Astronomy might provide the Medici a potent new set of emblems that associated their authority with celestial, and ultimately divine, sources. Discussions of natural and mechanical marvels might exemplify the value many early modern Continental courts placed on sparkling "civil conversation," the public display of wit and wonders, to amuse and amaze the prince and his courtiers. Throughout late Renaissance and seventeenth-century Europe "cabinets" of natural and artificial curiosities (fig. 20) were a notable feature of gentlemanly and aristocratic culture, where they figured as much in the "self-fashioning" of the socially ambitious as they did in systematic scientific inquiry. Moreover, governments going back to antiquity were well aware of possible military and economic uses of the mathematical sciences (fig. 24). The "militarization" of science is nothing qualitatively new in the twentieth century: practical studies of surveying and military fortification were important branches of "the mathematical sciences" in classical times; astronomy was always associated with the arts of navigation and long-distance political control, and it assumed even more importance in the great age of European expansion in the New World; the introduction of gunpowder meant that ballistics and metallurgy came to possess enormous significance to European states almost constantly at war with each other in the sixteenth and seventeenth centuries.

By the late sixteenth century humanistic currents—described in chapter 2—were also beginning to affect participation in science and the expectations that civil society might have of it. Humanists sought to purify knowledge by reexamining original Greek and Latin sources, and they stressed the importance of this reformed knowledge not just for professional scholars but for the activities of practically oriented civic gentlemen. As a partial result of humanist agitation (together with the invention of printing with movable type and the Protestant Reformation), the boundaries of literate culture began to be reshaped in the sixteenth and early seventeenth centuries. More and more *gentlemen* became avid consumers of a reformed body of knowledge. Practical ethical literature urged gentlemen to take up knowledge as an aid to virtue as well as civic utility. Universities that had been the almost exclusive preserve of an impoverished clergy and those training for a clerical role were increasingly urged by gentlemen to equip their sons with knowledge useful to civic life. And most influentially, writers close to the heart of European courts began to publicly urge the reform of learning, not just to suit it for the active lives of civic gentlemen but also to make learning a more effective arm of state power.

No writer was more enthusiastic or more influential than Francis Bacon, lord chancellor of England and court counselor to Queen Elizabeth I and King James I, in making a joint case for the reform of learning and the expansion of state power. In Bacon's view

blow
terra
A New
SYSTEM of MATHEMATICKS
Composed by the Eminently Worthy
Sr JONAS MOORE Knight
Late Surveyor Genl of his Maties
Ordnance and Fellow of the
Royall Society &c.

24. *The title page of* A New System of the Mathematicks *(1681) by Sir Jonas Moore (1617–79), surveyor-general of the ordnance in England and (sometimes unreliable) patron of science in the Royal Society of London. Mathematicians are depicted putting their knowledge to practical use, especially in surveying, navigation, and the measurement of time.*

the whole traditional body of learning needed to be reconstituted, but natural philosophy was meant to occupy a central role in this reform. The "distempers" of existing bodies of natural philosophy were visible signs that they were not genuine knowledge and not fit to contribute to the welfare of the state. "The first distemper of learning" was the practice of studying words rather than things. This was what led the Schoolmen to their "monstrous altercations and barking questions," and this was what needed to be reformed if natural philosophy was to become both credible and constructive: "It is not possible but that this quality of knowledge must fall under popular contempt, the people being apt to condemn truth upon occasion of controversies and altercations, and to think they are all out of their way which never meet." Indeed, a tendency to infer from the contentious status of a body of knowledge to the conclusion that none of it is reliable or true is probably characteristic of a wide variety of cultures, including our own.

Authoritarian states reckoned that matters of belief and its profession were their legitimate concerns. Individualism in belief, rather than being celebrated as a condition for intellectual progress, appeared to crown servants an object of anxiety. It was taken as a responsibility of the state, and the state church, to monitor and to manage belief in general, and when Bacon announced that he had "taken all knowledge to be my province," he was employing the Elizabethan English for the Latin *provincia*—an administrative district of the central government. Knowledge was to be effectively brought under the administrative competence of the state. Bacon was worried about centrifugal intellectual tendencies associated with the Protestant Reformation of the sixteenth century and especially its stress on individuals' competence to determine truth for themselves, by their own lights. He condemned intellectual individualists as "voluntaries," and later commentators denounced religious "enthusiasts" who claimed—without the mediation of priests—to know divine truth by direct inspiration.

To be sure, a measure of intellectual free action was the condition for reform—after all, the Schoolmen were criticized for their "slavery" to Aristotelian authority—but *uncontrolled* and *undisciplined* freedom of belief was deemed dangerous to good order. Accordingly, privacy and individualism in all sorts of knowledge threatened state power and authority, and Bacon's program for intellectual reform amounted to an attempt to secure order through means approved of, and implemented, by the state. As we have seen, *method* was prominent among these means. Method was represented as a machine for producing reliable and shared knowledge. But in Bacon's plan the implementation of proper method called not for disciplined individual reasoning (as it did for Descartes) but for organized *collective* labor. The reform of natural philosophy was to be accomplished by making the method-machine a tool of state bureaucracy. An enforced cure for the disorder of natural philosophy would remove a threat to the state and bring a potentially powerful source of intellectual authority within the compass of the state.

The utopian plan for this collective reform was developed in Bacon's *New Atlantis* (1627). Here Bacon described "Solomon's House" in the mythical land of Bensalem as a bureaucratically organized and intellectually differentiated research and engineering institute, serving the interests of an imperializing state. All the members of Solomon's House were state officials, on state salaries. And the purpose of their work was twofold: first, the extension of natural philosophy ("the knowledge of Causes") and second, the extension of power ("the enlarging of the bounds of Human Empire"). The work of Solomon's House powered the expansionist drive of the kingdom of Bensalem and received from the state, in turn, the resources to produce yet more knowledge.

Bacon had no doubt that a methodically reformed and disciplined natural philosophy could augment the power of those who controlled it. This was true in two senses. First, the control of knowledge was conceived as an instrument of state power. A state that abdicated its right to monitor what was believed was putting its authority at risk. Second, as Bacon famously said, "Human knowledge and human power meet in one." The ability of natural philosophical knowledge to yield practical outcomes and to produce the means for

25. *An imaginary visit by King Louis XIV* (center) *and his minister Colbert* (right) *in 1671 to the Académie Royale des Sciences. An air pump devised by the Dutch natural philosopher Christiaan Huygens (1622–95) is shown at the left. Source: Claude Perrault,* Mémoires pour servir à l'histoire naturelle des animaux *(1671); detail from an engraving by Sébastien Le Clerc.*

technological control of nature were taken as reliable tests of its truth. That was why a reformed philosophy was to have a legitimate claim on state treasure.

There were many differences between Bacon's mythical Solomon's House and the scientific societies and academies that began to appear about the middle of the seventeenth century—for example, the Florentine Accademia del Cimento in 1657, the Royal Society of London in 1660, and the Parisian Académie Royale des Sciences in 1666. Although all of them enjoyed a degree of princely and state patronage, only the Parisian society was at all well integrated with central government—its members received royal stipends, and royal treasure was lavishly expended on its scientific instruments—while

PROVANDO E
RIPROVANDO
Accademia del Cimento fiorita in Firen.ze sotto la protezione della
Real Casa dei Medici nel Secolo XVII

26. *A meeting of the Accademia del Cimento (Academy of Experiment), Florence, established after Galileo's death in 1642 by his followers Vincenzo Viviani and Evangelista Torricelli and their associates. Its patrons were two leading members of the ruling Tuscan Medici family, the Grand Duke Ferdinand II and Leopold, both amateur experimenters. This is an imaginary reconstruction of an academy meeting, in a 1773 engraving by G. Vascellini (in* Serie di ritratti d'uomini illustri toscani*). Some of the experimental instruments used by the Accademia are shown at the lower left. The bust on the wall is that of Leopold, and the Italian motto (* Provando e riprovando*) below the niche means "Try and try again," expressing members' commitment to experimental methods in natural philosophy.*

the London organization that advertised its Baconian inspiration expected much in the way of concrete royal support but received little more from the English Crown than the piece of parchment that chartered it. Nevertheless, there are several ways the emergence of the new scientific societies throughout Europe was a response to concerns for order similar to those that animated Bacon's writings.

First, the societies generally represented alternative organizational forms to the universities, and in many cases their leaders explicitly condemned hierarchical and disputatious universities as sites fit for a genuine natural philosophy. Bacon said that the "universities are the seat and continent" of the "distempers" of learning, while proponents of the Royal Society—many of them university men—identified the authoritarianism of the universities as inimicable to the progress of genuine knowledge. One early publicist wrote that "very mischievous consequences" for philosophy had flowed from the fact that "the Seats of Knowledge, have been for the most part heretofore, not *Laboratories,* as they ought to be; but only *Schools,* where some have *taught,* and all the rest subscribed." The universities, after all, were important institutions in forming the character of the young, and it was in this particular connection that their unreformed state might be accounted not just unfortunate but pernicious. The new societies aimed to provide a novel organizational form uniquely suited to the new practice; they made the production of new knowledge, rather than the just guardianship of and commentary on the old, central to their identity; and they aimed, with varying success, to link the progress of science to civic concerns rather than wholly scholarly or religious ones. Second, most of the new societies mobilized themselves around some more or less formal conception of method, and though their methodological allegiances differed, all of them placed high value on the necessity of disciplined collective labor in the making of proper natural knowledge. Individualistic sentiments remained strong in sectors of modern natural philosophy, but the very existence of scientific societies testifies how far reform was associated with intensely collective forms of activity.

Finally, the new societies manifested a pronounced concern for orderliness and the rules of proper behavior in making and evaluating natural knowledge: not for them the wrangling of traditional scholarly life. The legitimacy of the new knowledge was to be made manifest in the civility and good order of its collective production. A publicist of the early Royal Society of London announced that its membership was composed for the most part of "Gentlemen, free and unconfined," and, indeed, one marked contrast with traditional scholarly sites was the more civic, and more socially elevated, tone of several of the new societies. While Bacon made a humanist case for reforming natural philosophy so as to fit it for civic gentlemen, the participation in the Royal Society of men like the Honourable Robert Boyle—a wealthy and well-connected Anglo-Irish aristocrat—substantially transformed Baconian vision into social reality. The enterprise of natural knowledge was intentionally being made attractive to, and fit for, civic gentlemen. The consequences of changing participation in natural knowledge were considerable. A society dominated by gentlemen could more effectively draw on codes of genteel civility and decorum in conducting philosophical debate and evaluating testimony. Gentlemanly society had its own well-developed conventions for guaranteeing good order. The adhesion to natural philosophy of civic gentlemen thus offered a powerful alternative to scholarly disputatiousness.[1]

The codes regulating the "civil conversation" of early modern gentlemen warned against the intrusion of potentially divisive and disruptive topics. Ad hominem speech, as well as contentious matters of politics, theology, and metaphysics, was seen as threatening the good order and continuance of conversation. Just as the establish-

1. By no means all natural philosophers—even in the English Royal Society—*were* gentlemen. We still lack a secure understanding of the social map of scientific learning anywhere in Europe, and we do know that many important modern practitioners came from ungentle backgrounds. Nevertheless, the importance of gentlemanly codes of conduct in regulating behavior is formally independent of the identities of all the individuals operating under those codes. So, for example, knowledge of how to behave in church is not confined to the community of Christians, or even of believers in God. Nor was knowledge of how to behave as a gentleman restricted to those who were gentlemen.

ment of Boyle's matters of fact depended on protecting the boundaries between the factual and the theoretical, so the constitution of the Royal Society of London explicitly prohibited its fellows from speaking of religion or politics during the course of its scientific meetings, and similar prohibitions were inscribed in the charters of a number of Continental societies. A precursor to the French Académie Royale des Sciences, for example, announced its intention that "in the meetings, there will never be a discussion of the mysteries of religion or the affairs of state."[2] Such subjects, it was thought, could only divide people, and by the 1660s there was already some bitter experience with philosophical societies that split apart along grand metaphysical fault lines. Chapter 2 noted that many such matters were deemed inherently subjective, not amenable to rational treatment and rational agreement. The reformed natural philosophy was to offer its participants a quiet and orderly space from which an objective account of nature might credibly emerge and in which practitioners could civilly disagree without bringing down the whole house of knowledge.

Science as Religion's Handmaid

Late twentieth-century moderns are accustomed to hearing about the "inevitable opposition between science and religion," if, indeed, religion figures at all in our contemporary understanding of science and its history. Possibly much of what I have written in the preceding chapters about the mechanical philosophy, and about the relation between a reformed natural knowledge and secular concerns, has been read from that perspective. It is time to systematically correct any such impression, for the sense in which early modern changes in nat-

2. In practice, these prohibitions amounted to a ban only on *controversial* items of theology and politics. In societies whose members all took the existence of a creator God for granted, references, for example, to the divine origins of the world would not count as religious discussion, but allusions to the scope of human free will, or to the physical reality of transubstantiation, or to the proper relations between church and state might well be treated as controversial and divisive.

ural philosophy "threatened" religion or were animated by irreligious impulses needs to be very carefully qualified or even denied. In speaking about the purposes of changing natural knowledge in the seventeenth century, it is obligatory to treat its uses in *supporting* and *extending* broadly religious aims.

There was *no such thing* as a necessary seventeenth-century conflict between science and religion, but there were a number of quite specific problems for the relations between the views of some natural philosophers and the interests of some religious institutions that were precipitated by the changes treated in preceding chapters. From the medieval period Aristotelian natural philosophy had been "Christianized" in the culture of Scholasticism, and over a long period of adaptation, whatever mismatches there might originally have been between some "pagan" perspectives and Christian doctrine had been ironed out, reconciled, or simply set aside. The Roman Catholic Church not only had learned to live with the philosophies of ancient Greece and Rome, it had actively shaped some of them into systems of belief whose compatibility with Scripture and the doctrines of the church fathers was assumed. The institutions of Christian religion had evolved together with traditional bodies of natural knowledge, notably including those associated with Aristotle, Galen, and Ptolemy. This meant that any systematic challenge to traditional natural philosophy *might be* taken as an attack on elements of Christianity itself.

So, for example, Galileo's advocacy of Copernicanism as a physically true account of the cosmos was applauded by some quarters of the Catholic Church but eventually encountered vigorous opposition from the Inquisition. In denying the geocentric and geostatic system of Ptolemy, Galileo was taken as rejecting the truth of Scripture. The Bible did indeed make periodic references to the stability of the earth and the movement of the sun, and Galileo accounted it both "very pious to say and prudent to affirm that the holy Bible can never speak untruth." The reference in the Book of Joshua to the sun standing still was to be accepted as true. But here Galileo insisted on *at least* the equal status of God's Book of Nature as a source of truth and, consequently, on *at least* the equal status of natural philosophers as experts in interpreting divinely inspired texts. Copernicanism was said to be physically true according to the best interpreters of the Book of Nature, and since it was agreed that two truths cannot contradict each other, Galileo maintained that biblical references to the stability of the earth and the mobility of the sun were to be taken not as *literal* truths but as *metaphorical* devices, adapting articles of faith to vulgar capacities, "lest the shallow minds of the common people should become confused." This strategy sought to open up a legitimate space for the expertise of the natural philosopher, independent of theological expertise, but not in any necessary conflict with it. The Book of Nature that was the source of the natural philosopher's expert knowledge was, after all, just as much a source of divine knowledge as Scripture. In fact, Galileo arguably wanted more than cultural equality for the natural philosopher: he intermittently contrasted the ambiguity of scriptural texts with the interpretive clarity of the Book of Nature. This was a sense in which the expert natural philosopher might be understood as doing a *better* job of interpreting God's word than the theologian.

That strategy was not locally successful, and the outcome of Galileo's famous trial by the Roman Catholic Church in 1633 was the requirement that he abjure professing the *physical truth* of Copernicanism.[3] Yet elsewhere aggressive practitioners in more receptive settings vigorously and clearly claimed not just that the new natural philosophy ought to be conceded an independent expertise and basis for credibility but also that science could provide uniquely potent resources to support and extend Christian religious belief. In fact the justification of the new practice through claims of religious utility was, in all European settings, an important resource for achieving cultural legitimacy. This was a deeply religious age, and religious in-

3. The difference between advancing Copernicanism as a mathematical predictive model and claiming that it was a truthful physical account is not usually appreciated in popular stories about Galileo's trial. His Catholic critics saw few problems with the former and were mainly troubled by the latter. As we have already seen, natural philosophers not vexed by the church debated similar issues bearing on the identity of the practices known respectively as "natural philosophy" and "mathematics."

stitutions in all European countries exercised enormous secular power, both in their own right and as associates of the state. No new strand of culture that was widely seen as threatening religion could hope to become institutionalized.[4]

In Protestant England, advocates of a reformed natural knowledge argued that a proper reading of the Book of Nature could support Christian religion by purifying it. Over the centuries superstitions and unauthenticated fables had illegitimately been attached to religion, most especially in its papist form: "Narrations of miracles wrought by martyrs, hermits, or monks of the desert, and other holy men, and their relics, shrines, chapels, and images." These bogus claims, Bacon wrote, were all liable to be exposed "as old wives' fables, impostures of the clergy, illusions of spirits, and badges of antichrist, to the scandal and detriment of religion." The techniques of intellectual quality control recommended for a reformed natural history could be used to winnow out testimonial wheat from chaff, to purge Protestant Christianity of idolatrous elements and restore it to its primitive purity. Bacon agreed with Galileo that Scripture was a book demanding expert interpretation if its true meaning was to be discerned. But if the parallel Book of Nature could be read aright—with the discipline of proper method—then the natural philosopher could contribute as much as the theologian, if not more, in establishing religious truth and in ensuring right belief. To be sure, science and theology might be identified as distinct enterprises—as we have seen in the case of the mechanical philosophy and its cultural boundaries—but it was this very separation that allowed a reformed natural philosophy to contribute *independently* to religious concerns.

4. This general view of the relation between legitimacy and institutionalization was articulated in 1905 by the sociologist Max Weber, and in 1938 another sociologist—Robert K. Merton—developed a celebrated thesis about the positive relations between science, technology, and religious culture in seventeenth-century England. Much of what Merton then wrote about religious motives to science, and religious justifications for science, has passed into historical commonplace, while some related claims about the precise religious affiliations of English men of science remain highly controversial.

It was true that the natural philosopher's role was conceived mainly as dealing with what were called "secondary" or "efficient causes," that *by which* an effect is brought about—for example, identifying the movement of one material body as the efficient cause of another's movement. And it was acknowledged to be true that a superficial orientation toward such causes might lead practitioners to ignore "final causes," the ultimate causes of movement or that *for the sake of which* movement occurred. But if superficial natural knowledge disposed people toward denying God, then a proper and profound natural philosophy offered solid assurance of God's existence and attributes. That was the sense of Bacon's claim that "natural philosophy is after the word of God at once the surest medicine against superstition, and the most approved nourishment for faith, and therefore she is rightly given to religion as her most faithful handmaid." It was considered a religious duty that human beings use their God-given faculties of observation and reason to read the Book of Nature and to read it properly.

Just as a reformed natural philosophy might help restore a pure and primitive Christianity, so the technical utility of the new practice might restore humanity to its rightful dominion over nature. When Bacon wrote *The Great Instauration,* he signaled his (widely shared) conviction that humanity had, through the fall from grace in the Garden of Eden, lost its original technological control over nature. It was also a religious duty to restore that sovereignty, and the new natural philosophy was meant to be a powerful tool in that task. A Baconian imperative toward the control and reshaping of nature was realized in the early Royal Society of London, where one of its original institutional projects consisted of collecting the previously uncodified knowledge of trades and crafts, passing that craft knowledge through a sieve of philosophical scrutiny, and then attempting to restore an improved, more useful version to domains of practical activity. That enterprise was described and justified in a religious idiom. At the middle of the seventeenth century some practitioners even saw the restoration of technological control in millenarian terms: only when humanity had by its own efforts restored its original dominion over nature would Christ come again, to rule on earth for a millennium—a thousand years—before the general resurrection. That was what had been prophesied in the Book of Daniel.

Convictions that a reformed body of natural knowledge would achieve technically useful results were vigorously expressed by both English and Continental practitioners. And whatever the validity of utilitarian claims, the *promise* of utility is undoubtedly pertinent to accounting for the attractiveness of much of the new practice. Reformed knowledge, especially in its mechanical modes, was to be as technologically fertile as the Scholastic alternative was evidently barren. Use was to be a reliable test of truth. If Bacon's vision was in this respect the most aggressively optimistic, the general promise strongly linking reformed knowledge to unique utility was pervasive. In France, for example, Descartes was as sure that contemporary medicine was ineffective as he was convinced that a proper causal knowledge of the body (on mechanical principles) would aid in preserving health and prolonging life: "We could be free of an infinitude of maladies both of body and mind, and even possibly of the infirmities of age, if we had sufficient knowledge of their causes." (Descartes's views on philosophically reformed medicine were so well known that when he died at age fifty-four—of a chill contracted on a freezing Swedish morning—one of his friends insisted that without "an external and violent cause" he would have lived five hundred years!) In England, Robert Hooke promised no end of useful outcomes if only the true causal structure of nature was made known and the proper method of discovery was employed: Why not the transmutation of base metals into gold? Why not the art of flying?

The question of the real historical relation between the growth of scientific knowledge and the extension of technological control has been endlessly debated by historians and economists. On the one hand, it now appears unlikely that the "high theory" of the Scientific Revolution had any substantial direct effect on economically useful technology in either the seventeenth century or the eighteenth. Utilitarian motives among many modern natural philosophers do not automatically equate with substantial economic consequences, and many seventeenth-century commentators found the new philosophers' promises not just false but funny. On the other hand, we have already noted intimate links between the "mixed" (or "impure") mathematical sciences and military and productive technology going back to antiquity, and there is no reason to think that such links were not strengthened through the early modern period. Moreover, there can be little doubt that the vast expansion of natural historical and geographical knowledge that attended the voyages of exploration and conquest contributed significantly to the making of empires and fortunes. It is the link between "theory" as a cause and technical change as an effect that remains at issue.

The possible effect of economic concerns on changes in scientific knowledge has also been debated at length. In the 1930s Robert Merton famously claimed to have showed a clustering of scientific work of the early Royal Society in areas of potential economic or military application, arguing that these "foci of interest" were evidence of the influence of wider social concerns on the dynamics of science. Here again the operative word is "potential," since historians have had great difficulty in establishing that any of these spheres of technologically or economically inspired science bore substantial fruit. Baconian rhetoric, that is to say, translated poorly into practical reality, and the military-industrial-scientific complex is more properly regarded as a creation of the nineteenth and twentieth centuries. It is, however, one thing to look for the uses of scientific *knowledge* and another to consider the spheres of practical activity engaged in by scientifically trained *people.* Scientifically derived *information, skills,* and perhaps *attitudes* were important resources in all sorts of practical activities, and there is no problem in identifying many seventeenth-century natural philosophers and natural historians who used these resources in economically and militarily consequential ways. Marxist historians have made particularly valuable contributions toward understanding how closely reforms in natural knowledge were associated with new social and cultural relations between "scholars" and "craftsmen," and between natural philosophical changes and changes in the economic and political orders. If the new natural philosopher was rarely himself a "craftsman," still he was far more likely to have knowledge of craft and productive concerns than was the traditional practitioner.

Nature and God, Wisdom and Will

To present-day sensibilities it is the mechanical aspect of the new natural philosophy that must appear most seriously at odds with religious belief. If nature is a great machine, then what need of God or even of spiritual agencies to understand how nature works? Yet it was precisely the mechanical conception of nature that generated some of the most powerful and persuasive arguments that the new practice was religion's truest handmaid. Just because machines were conceived of as impersonal—their characteristics to be juxtaposed to the intelligent and purposeful life of human beings—a mechanical metaphor for nature posed questions about the *apparent evidence* in nature of intelligence and purpose. How was it that, if nature was really a great machine, one was to explain the appearance of complex patterns, vitality, and purposiveness? Put another way, how ought a mechanical philosophy to deal with those aspects of nature to which traditional organicist and animist philosophies responded so strongly?

That nature showed solid evidence of design—that it was artfully contrived—was wholly accepted by mechanical philosophers. But if that design was not to be accounted for by the indwelling intelligence of material nature, then artful contrivance had to arise from something outside nature itself. This train of inference was the basis of the most pervasive seventeenth-century argument for the existence and intelligence of a deity—the *argument from design*—which linked the practice of science to religious values from the early modern period through the nineteenth century.[5] The clock metaphor

5. The argument from design was the cornerstone of *natural theology,* that is, the practice of establishing the existence and attributes of God from the evidence of nature. It was the premises and reasoning of the argument from design that Charles

again. Imagine that one is walking along a road and finds a watch lying on the ground. Taking it apart, one observes how intricately its mechanical parts are put together and how well adapted they are to the evident function of the watch in telling time. In just the same way, those who observed and reflected on the natural world were confronted with the solid evidence of design and the inescapable conclusion that there was an intelligent designer, one whose intelligence was unimaginably greater than that of the human artificer.

So Boyle wrote of the material parts of the human body as mechanical contrivances. And when the mechanically informed anatomist "has learned the structure, use, and harmony of the parts of the body, he is able to discern that matchless engine to be admirably contrived, in order to the exercise of all the motions and functions, whereto it was designed: and yet [this anatomist], had he never contemplated a human body, could never have imagined or designed an engine of no greater bulk, any thing near so fitted to perform all that variety of actions we daily see performed either in or by a human body." The more we learn about the world-engine, the more we are persuaded not just of the existence of a creator God but also of his creative wisdom. No such engine could conceivably have come into existence by the chance concurrence of corpuscles. In the 1670s the French Cartesian Nicolas Malebranche (1638–1715) agreed: "When I see a watch, I have reason to conclude, that there is some Intelligent Being, since it is impossible for chance and haphazard to produce, to range and position all its wheels. How then could it be possible, that chance, and a confused jumble of atoms, should be capable of ranging in all men and animals, such abundance of different secret springs and engines, with that exactness and proportion?" This clear evidence of contrivance in the natural world was, as Boyle said, "one of the great motives" to religious belief, and those whose natural knowledge was greatest were said to be most disposed to venerate God's creative wisdom. In 1691 the English naturalist and divine John Ray (1627–1705)

Darwin's mid-nineteenth-century materialist account of evolution by natural selection was directed against.

offered the animal eye as a powerful exemplar of God's designing intelligence and beneficence, as well as of the certainty with which humans could come to be assured of his existence and attributes:

> That the Eye is employed by Man and all Animals for the use of Vision, which, as they are framed, is so necessary for them, that they could not live without it; and God Almighty knew that it would be so; and seeing it is so admirably fitted and adapted to this use, that all the Wit and Art of men and Angels could not have contrived it better, if so well; it must needs be highly absurd and unreasonable to affirm, either that it was not designed at all for this use, or that it is impossible for man to know whether it was or not.

The new optical instruments that vastly expanded "the empire of sense" were likewise said to be active encouragements to religious belief. On the one hand, as chapter 1 indicated, the microscope gave confidence to those advocating a corpuscular structure of matter: the granular or jagged surfaces it revealed (fig. 11) in apparently smooth and homogeneous bodies offered a token of what might in time come to be revealed at an ultimately small level. On the other hand, microscopically enhanced vision displayed worlds of hitherto unsuspected complexity, beauty, and contrivance in what had been regarded as the most "insignificant" and "despicable" creatures. Under magnification, the eye of the common fly was shown to be a wonderfully contrived optical device, superbly adapted to the total structure of the fly's body and to its way of life (fig. 27). Everything in God's created nature displayed his power, goodness, and wisdom. The relation between structure and function revealed by the microscope was of such adaptive excellence that it would be, as Hooke wrote, "impossible for all the reason in the world to do the same thing that should have more convenient properties." Who would be so stupid "as to think all those things the production of chance?" Either their reason "must be extremely depraved, or they did never attentively consider and contemplate the Works of the Almighty." Moreover, both microscope and

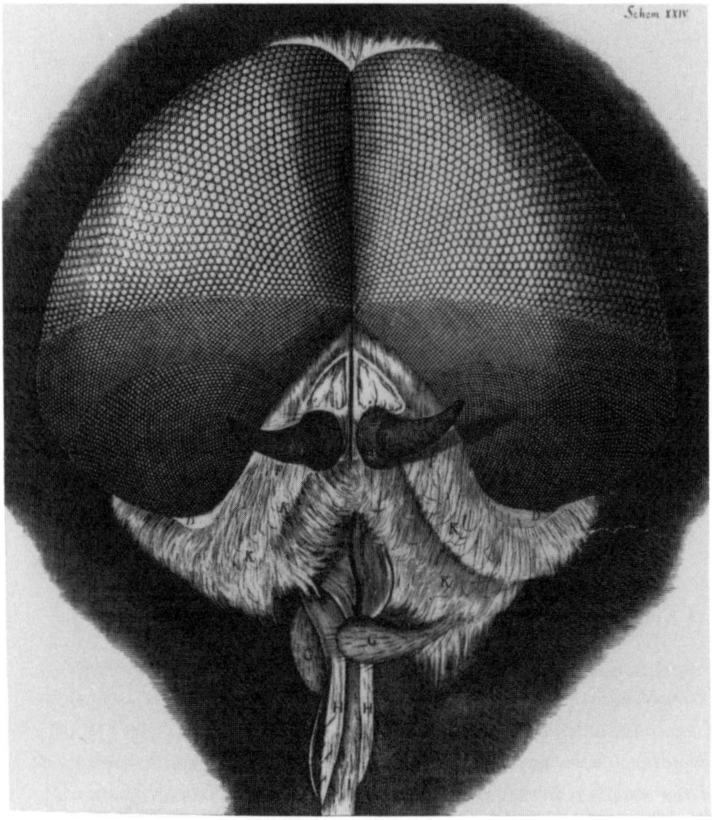

27. *The eyes of a common fly microscopically magnified by Robert Hooke. Hooke counted about fourteen thousand elements (or "pearls") in these eyes and did not doubt "but that there may be as much curiosity of contrivance and structure in every one of these Pearls, as in the eye of a Whale or Elephant, and that the almighty's* Fiat *could as easily cause the existence of the one as the other; and as one day and a thousand years are the same with him, so may one eye and ten thousand." Source: Robert Hooke,* Micrographia *(1665).*

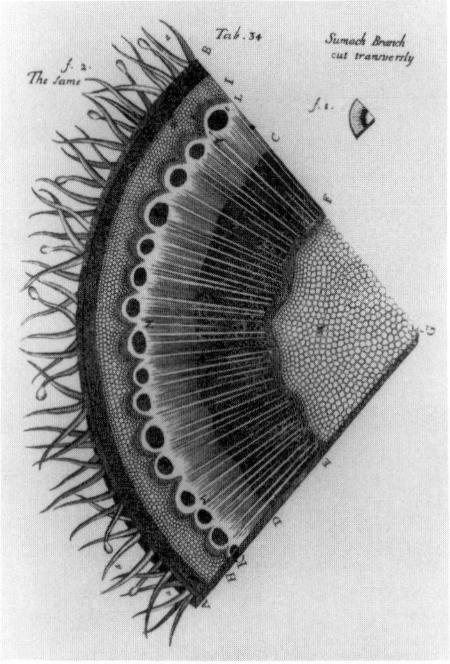

28. *A section of a sumac stem as depicted by the English naturalist Nehemiah Grew (1641–1712). Grew here shows both an unmagnified and a microscopically magnified quadrant of a stem of the common sumac tree (*Rhus*). Note the detailed representation of vessels whose physiological functions Grew sought to identify (partly by analogy with better-understood animal structures). These features can be seen without a microscope, but much more detail is revealed microscopically. Grew was concerned to show what structures various plants had in common and in what ways they were specifically differentiated, both serving to display "the Constant and Universal Design of* Nature.*" His observations were similar to those made slightly earlier by the Italian Marcello Malpighi (1628–94). Grew was then secretary of Royal Society of London, and his book was sponsored by the Society and dedicated to its patron King Charles II. The dedicatory epistle celebrated the complex design microscopically observable in even the most common natural objects: "One who walks about with the meanest Stick, holds a Piece of Nature's Handicraft, which far surpasses the most elaborate Woof or Needle-Work in the World." Source: Nehemiah Grew,* The Anatomy of Plants *(1682).*

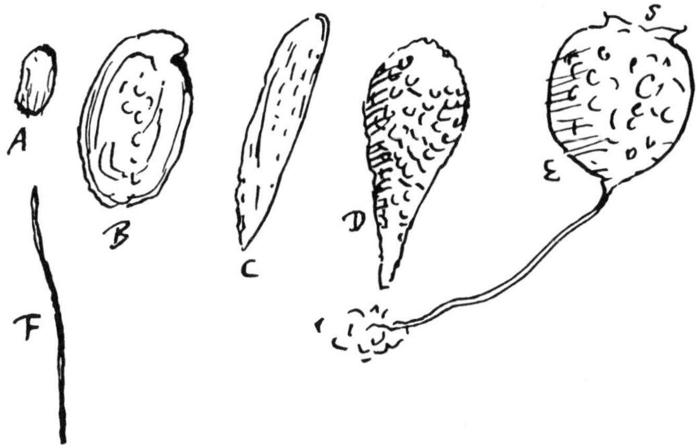

29. *Antoni van Leeuwenhoek's observations of protozoa. The drawings here represent observations by Christiaan Huygens in 1678, sent to Leeuwenhoek for confirmation by way of Christiaan's brother Constantine. Christiaan had originally been skeptical of Leeuwenhoek's claims. Leeuwenhoek told Constantine that these drawings were probably reliable representations of the same sorts of "animalcules" he himself had seen and reported three years earlier. Twentieth-century scientists reckon that D and E may have been forms of the protozoa* Stentor *and* Vorticella *respectively. Source: Letter from Christiaan Huygens to Constantine Huygens, 18 November 1678, in* Oeuvres complètes de Christiaan Huygens, *8:124 (partial translation in* The Collected Letters of Antoni van Leeuwenhoek, *2:399–407).*

telescope testified to the hitherto unknown *range,* as well as the beautiful contrivance, of God's creation: Why not believe in the dense inhabitation of the planets when, as Leeuwenhoek showed, a drop of water may seethe with small living beings (fig. 29)? In the 1680s the French philosopher Bernard de Fontenelle (1657–1757) wrote, "We see from the elephant down to the mite; there our sight ends. But beyond the mite an infinite multitude of animals begins for which the mite is an elephant, and which can't be perceived with ordinary eyesight." The lens thus opened up a new domain of wonder and a new inducement to belief. The biblical psalmist sang of the way "the heavens declare the glory of God; and the firmament showeth his handiwork," but the modern microscope and telescope showed even more glory and even more wisdom.

A mechanical conception of nature could support belief in the existence of God at the most fundamental levels. We are to understand that a watch with moving hands and gears was at some point *set in motion*. We are also to accept that its component parts do not move themselves and are utterly dependent on an external motive agency. All this follows from the inanimate conception of matter described in chapter 1. However intricately its parts are formed and adapted for the purpose of telling time, the watch cannot actually perform that function until its mechanical elements are externally animated. So if we are to accept the physical legitimacy of a mechanical conception of nature, then all the pervasive evidence of motion in the world is testimony to the animating work of a creator deity. It was just this sensibility that a wide range of seventeenth-century mechanical philosophers, especially but by no means exclusively in England, worked to encourage. Matter cannot move itself. It can be moved by contact with another bit of moving matter, but ultimately movement has to have an origin that is itself not material. That was its final cause, and many mechanical philosophers maintained that a mechanical understanding of nature would lead us to the recognition of a final cause that was itself not natural but supernatural, not material but spiritual. The proper study of nature led "from Nature up to Nature's God."

Some mechanical philosophers were content to posit a creator God who wound up the world-clock at the moment it was made, after which it could be conceived to run perfectly, a testament to the wisdom of a deity whose creation was so flawless that it needed no further tinkering or superintendence. Religiously concerned English mechanical philosophers worried that Descartes—who formally banished talk of final causes from his natural philosophy—might be taken as supporting such a conception of God's relation to nature: "We should not," Descartes wrote, "be so arrogant as to suppose that we can share in God's plans." And though the French philosopher intermittently gestured at a more active divine role, many English practitioners anxiously debated whether Cartesian philosophy made that role as evident as the effective defense of Christianity required. Later in the seventeenth century a group of turbulent "deists" throughout Western Europe sought to restrict God's role in nature to his creative act and his attributes to wisdom in creating a perfectly running world-machine. English mechanical philosophers from Boyle to Newton were not content with such a conception of God, arguing that it was philosophically incorrect as well as theologically unsafe.

The object of their opposition was a conception of God as the world's "absentee landlord," a deity who created the world and then meddled no further in its affairs, natural or political. Such a notion of God was considered to threaten important articles of proper Christian faith and vital aspects of moral order. The miracles related in both Old and New Testaments were held to be central proofs of the Christian religion. They involved God's active momentary intercession in the world—what was called his "special" or "extraordinary providence." Modern natural philosophers debated whether "the age of miracles" was confined to the biblical past and was now over. Some—like Hobbes—said that it was; others—like Boyle—were not so sure; while many Continental Catholic practitioners—like Mersenne and Pascal—were convinced that modern miracles did occur and occupied themselves with techniques for authenticating miracle claims. Yet most English philosophers, at any rate, were keen not to take any view that suggested limits to God's power to do as he liked in the natural order. A correct natural philosophy, accordingly, allowed scope for God's intermittent exercise of divine *will* in the world as well as encouraging recognition of his creative *wisdom*.

This "voluntarist" strand was, again, highly developed in English natural philosophy from Boyle to Newton. Nature was to be inspected for the evidence of regularity and patterns that testified to God's designing intelligence. Such evidence spoke to God's "general" or "ordinary providence," and it was to be concluded that natural regularities were continuously and actively maintained by God. The laws of motion, Boyle wrote, were "freely established and still maintained by God." But a properly conceived study of nature was also to be a resource for showing people that God constantly monitored, superintended, and intervened in the world. And that conception of God's role was as much a social as a theological instrument, for, it was thought, only if people were aware that God "was watching" would right conduct have its ultimately powerful motive. So Boyle, for example, was uncomfortable with certain common understandings of the "laws of nature" and repeatedly cautioned that such usages should be avoided or at least carefully qualified. Regularities were to be observed in nature and might even be mathematically expressed. Yet it was to be understood that all such regularities were subject to God's pleasure. At any moment he might exercise his power—his special providence—to change the inertial disposition of matter, to suspend or alter its usual behavior. And Scripture truthfully related instances in which he undoubtedly had done so. It should be understood that stones fell downward, at thirty-two feet per second squared, *God willing*.

Chapter 1 mentioned Father Marin Mersenne's worries about the dangerous religious and moral consequences of "Renaissance naturalism," the body of thought that attributed inherent activity, and even sentience, to nature itself. Throughout his career, Boyle shared similar anxieties about such tendencies. Acknowledging the "brute stupidity" of matter, and the animating power of an external spiritual agency, was seen as a condition for building both a proper natural philosophy and a secure Christian moral order. Consider the phenomenon of suction. A person sucks on a straw, and water rises up from the vessel to the mouth. As we have seen, a traditional explanation attributed to the water a fear or abhorrence of the vacuum whose formation was threatened by withdrawing air from the top of the straw. In contrast, the mechanical philosophy might interpret the water's rise as an effect of differential pressures or weights of the air bearing on the liquid in the straw and on the vessel below. There was no philosophical need to attribute anything like purpose or sentience to matter.

Boyle elaborated his view of suction in a number of tracts written during the middle decades of the seventeenth century and printed late in his life. Here he was concerned to refute not only the Aristotelian *horror vacui* but also a more widely distributed "vulgar notion of nature." The "true physical causes" of such effects would never have been discovered "if the moderns had acquiesced, as their predecessors did, in that imaginary one, that the world was governed by a watchful being, called nature, and that she abhors a vacuum, and consequently is still in a readiness to do irresistibly whatever is necessary to prevent it." The notion that nature abhors a vacuum supposes "that a brute and inanimate creature, as water, not only had a power to move its heavy body upwards . . . , but knows both that air has been sucked out of the reed . . . and that this water is withal so generous, as by ascending, to act contrary to its particular inclination for the general good of the universe, like a noble patriot, that sacrifices his private interests to the public ones of his country."

The philosophical arguments against attributing purpose to matter have been sketched in preceding chapters. Here Boyle contended that this "vulgar notion" was also noxious to true religion and to the moral order that religion supported: it was "dangerous to religion in general, and consequently to the Christian." The problem centered on how much should be attributed to nature and how much to God, rightly considered as an external spiritual agency. Because "many atheists ascribe so much to nature, . . . they think it needless to have recourse to a Deity for the giving an account of the phenomena of the universe." If you ascribe activity and intelligence to what is properly conceived as brute nature, you encourage the belief that material nature is self-sufficient, not dependent on external animating agencies for its motions and patterns. Then you make nature a "semideity" while derogating God's majesty and power. The "vulgar notion" of nature had been, and continued to be, a great cause of idolatry and atheism: "The looking on things merely corporeal, and oftentimes inanimate things, as if they were endowed with life, sense, and understanding; and the ascribing to nature [capacities] that belong but to God, have been some . . . of the grand causes of . . . polytheism and idolatry."

And to be sure, during the English civil wars and Interregnum of the mid-1640s to the late 1650s, there was a great flourishing of radical political thought, some strands of which used "the vulgar notion of nature" to argue against a range of religious and political hierarchies. What need of an external animating deity if activity was distributed throughout material nature? What need of a priestly caste to interpret God's ways to people if spiritual powers were distributed through all things and within all believers? Such English radical sects as the Diggers and Ranters argued vehemently for an animistic view of nature as a tool in their political programs: God—the source of all activity and purpose—was "in all things," in "man and beast, fish and fowl, and every green thing, from the highest cedar to the ivy on the wall." And if soul-like properties were immanent in matter, why accept the crucial orthodox Christian doctrine of the soul's afterlife, where it was punished or rewarded according to its merits? Might it not be said that the soul died with the body? Boyle opposed such views of nature and their moral consequences, and in so doing he showed how moral as well as technical considerations figured in arguing the legitimacy of a mechanical philosophy.[6]

By the 1680s Isaac Newton's celestial physics appeared to offer scientific proof of divine intervention in nature. Newton's calculations seemed to show that the solar system had a tendency, over time, to collapse in on itself. A periodic "reformation" of solar systemic order was required, and Newton maintained that such reformations had occurred, as evinced by the system's continued existence. It might be that God used natural agents to effect this periodic reformation—Newton speculated about the role of comets in this respect—or it might be that he intervened directly. Either way, a voluntarist conception of God's activity in nature was built into the heart of the Newtonian system. It was considered not an imperfection but a

6. Although the radical sects were effectively crushed by the restoration of the English monarchy in 1660, broadly similar cultural tendencies (including the "deism" mentioned above) erupted again in the late seventeenth and early eighteenth centuries, when they were countered by philosophical arguments comparable to Boyle's.

recommendation of the mechanical system of mathematical natural philosophy that God's intercession was required by it and publicly manifested through it.

Mechanical philosophers varied importantly in their attachment to recognized religious goals. Hobbes and Descartes professed religious commitments and argued for the compatibility of their practices with broad clerical ends. For their pains their professions were widely disbelieved, and though atheism—in the sense of a formally expressed conviction that there was no God—probably did not exist in seventeenth-century literate culture, religiously inclined practitioners worried that the philosophies of both Hobbes and Descartes accorded God such a limited role in the world that they provided aid and comfort to popular atheism. The dominant English tradition, however, took it for granted that a key function of natural philosophy was to support and enhance Christian religion. The role of the natural philosopher thus overlapped massively with that of the cleric; as Newton said, "To discourse of [God] from the appearances of things does certainly belong to natural philosophy."

Just as priests were traditionally defined by their authority to interpret Scripture, so many religiously inclined natural philosophers considered themselves to be, in Boyle's term, "priests of nature," possessing expert ability to interpret the Book of Nature and to make it available for religious uses. They were charged with producing "successful arguments to convince men there is a God" possessing the attributes of wisdom and power. In 1661 Robert Hooke wrote that experimental natural philosophy was "certainly the most likely way to erect a glorious and everlasting structure and temple to nature," and thus to nature's creator. And the Cambridge Platonist philosopher Henry More (1614–87) lauded the "more perfect Philosophy" of the Royal Society, "which is so far from tending to Atheism, that I am confident it will utterly rout it." Boyle described experimental research as a kind of worship: it was therefore fit that laboratory work be performed, like divine service, on Sundays. The English mechanical philosopher was represented as a godly man, fit to celebrate divine service in the temple of nature.

It was the godly identity of these mechanical philosophers that powerfully equipped them for the task of displaying the *real action* of spirit in the world. Their role was to offer material and mechanical explanations where they could and, where they could not, to acknowledge the limits of mechanism. Just as the idea of inanimate mechanism generated the complementary category of spiritual agency, so the notion of a mechanically operating nature was held to generate that of supernatural powers working in and on nature. Displaying the *limits* of mechanism was not seen as a defeat for a correctly conceived natural philosophy. Were seventeenth-century mechanical philosophy regarded as a wholly secular enterprise, then such limitations might count as a defeat, but it was not wholly secular. The scope of mechanism, as such, might have limits, but the beliefs of mechanical philosophers were not to be confined to what could be mechanically explained. Boyle said that the earth and the air above it "are frequented by multitudes of spirits" and that God had created "an inestimable multitude of spiritual beings, of various kinds, each of them endowed with an intellect and will of its own." Chapter 2 noted that many leading fellows of the Royal Society professed belief in demons and witches, and Boyle wrote that such beliefs were theologically useful: "To grant . . . that there are intelligent beings that are not ordinarily visible does much conduce to the reclaiming . . . of atheists"; they would "help to enlarge the somewhat too narrow conceptions men are wont to have of the amplitude of the works of God." What the mechanical philosophers believed *as* mechanical philosophers was not coextensive with what they might legitimately believe was true about the world.

Moreover, some mechanical philosophers undertook to verify spirit testimonies—to sort out those that might have a natural explanation from those that did not—in order to reconstitute our supernatural knowledge on a firmer foundation. Testimony about miracles and the actions of spirit had to be vigilantly policed. Uncontrolled reporting of spiritual actions, and uncontrolled belief in miracles, worked to subvert legitimate authority and to corrupt religion. Again, private belief might be socially dangerous. If uninstructed individuals believed just what they liked, with no external authority to judge whether the beliefs were valid, then disorder would be the consequence. The experimental community, however, had shown itself able to do the job of intellectual policing effectively. Its techniques for establishing matters of fact had given visible proof of its orderliness and its disinterestedness. In 1668 Joseph Glanvill wrote, "We know not anything of the world we live in but by *experiment,* and the *Phenomena;* and there is the same way of *speculating immaterial* nature." Mechanical philosophers might call for, and help to produce, *authenticated* accounts of spiritual phenomena. The handmaid to religion could help with the task of constituting the stock of genuine religious knowledge.

Nature and Purpose: The Place of Mystery in a World of Science

A theme running throughout this account of the Scientific Revolution has been modern practitioners' suspicion, and even vigorous rejection, of teleological accounts of natural phenomena—that is, of explanations that identified the purpose of natural effects as their cause. The theme runs from Galileo's and Hobbes's critique of Aristotelian "natural place" doctrines to Mersenne's mechanical replacement of "Renaissance naturalism" to Boyle's wariness about the philosophical use of the language of "natural laws." A simplistic summary might therefore be tempted to conclude that this theme captures the "essence" of the Scientific Revolution, or at very least the essence of the mechanical philosophy: mechanical explanations just replaced teleological explanations, and so modernity was made.

Yet the preceding section has just shown that such a global conclusion would be incorrect. Very many seventeenth-century practitioners reckoned that the scope of mechanical explanations was limited. To speak *as* a natural philosopher might be to speak *in mechanical terms,* setting aside notions of purpose, but even so the scope of what could be mechanically explained might not be deemed identical with the scope of the world's phenomena. And other practitioners, equally persuaded of the explanatory value of mechanism, accepted no such limitations on what kinds of accounts the natural philosopher might properly bring to bear on the world. On the one hand, Descartes proceeded by imagining a hypothetical natural world that God *might* have created, a world wholly amenable to mechanical explanation: this was the world the natural philosopher was to explain. On the other hand, such writers as Robert Boyle and John Ray were concerned to trace the evidence of God's purpose and design in the world he *did* create. That is why they were comfortable with the philosophical propriety of giving explanations in terms of purpose when, as they reckoned, the evidence of nature unambiguously supported such conclusions. The argument from design that constituted the keystone of "natural theology" was in this sense a teleological explanation: it explained the adaptation of natural structure to function in terms of divine purpose. These differences in explanatory strategies reflect different conceptions of the proper business of natural philosophers and natural historians. All practitioners might agree in principle that a reformed understanding of nature should allay doubt, secure right belief, and ensure the adequate foundations for moral order, yet they diverged in their notions of how natural inquiry might be framed to best fit it for those tasks.

For some philosophers there was to be a proper role for non-mechanical and teleological explanations in the understanding of nature. In such cases one should not speak of an "incomplete" mechanism, for that might imply that the business of a natural philosopher was just to give mechanical explanations, regardless of the nature of the phenomena and of the evidence available to support such explanations. Boyle, for example, offered a measured defense of the philosophical propriety of final causes, most especially, but not exclusively, in the explanation of living things: "There is no part of nature known to us wherein the consideration of final causes may so justly take place, as in the structure of the bodies of animals." The sheer complexity of animate structures, as well as their evident adaptation of structure to function, specially impelled belief in the "superintendency of a rational agent." One might say that such complexity and adaptation were brought about by mechanical means, but not that they were brought about without intelligent design.

By the end of the seventeenth century and the early eighteenth Sir Isaac Newton, the philosopher who is sometimes celebrated as bringing the mechanical philosophy to perfection, was expressing impatience at global philosophical attempts to give mechanical explanations. Newton's reluctance to specify a mechanical cause for gravitation was noted in chapter 1, but similar limits to mechanism also informed his treatments of, for example, magnetism, electricity, and the phenomena of life. By what right was the supplying of specifically causal mechanical accounts to define the business of natural philosophy? Why *not* talk of attractive and repulsive forces, of the "active powers" in nature if the phenomena manifested them? To do so was not necessarily to speak either as an Aristotelian or as a Renaissance occultist: "These principles I consider, not as occult Qualities, supposed to result from the specific Forms of Things, but as general Laws of Nature, by which the Things themselves are formed; their Truth appearing to us by Phenomena, though their Causes be not yet discovered. For these are manifest Qualities, and their Causes only are occult." It was not philosophical, but its opposite, to "feign" (or imaginatively concoct) hypotheses, even and especially mechanical causal hypotheses, when the senses and the intellect could not securely discover them. "In bodies," Newton wrote, "we see only their figures and their colours, we hear only the sounds, we touch only their outward surfaces, we smell only the smells, and taste the savours; but their inward substances are not to be known either by our senses, or by any reflex act of our minds." If other practitioners equated proper natural philosophy with the provision of mechanical accounts, Newton here professed contentment with the ultimate inscrutability of nature. The demand for intelligibility had to be taught its proper limits. Natural philosophy was to be a solid rock of certainty, lapped on all sides by vast seas of mystery.

The preceding discussion intimated that among seventeenth-century moderns René Descartes was the most unremittingly aggressive and optimistic supplier of causal mechanical explanations. So he was, and in contrast with such philosophers as Boyle and Newton, nowhere was Descartes's mechanical ambition more thoroughly realized than in his accounts of *animate* phenomena, including the workings of the human body. It is in this connection that one can best appreciate both the limits of mechanical accounts and the consequences of mechanical frameworks for understanding the place of human beings and human experience in the modern conception of nature.

Chapter 1 briefly introduced Descartes's account of the human body as a "statue, an earthen machine." He meant to bring the workings of the human body wholly within the scope of a mechanical philosophy: the digestion, assimilation, and excretion of food; the formation of blood and its movement through the veins and arteries; the actions and functions of respiration; and the patterns of "reflex action" depicted in figure 10.[7] The machine that constituted the human body was, to be sure, "incomparably better arranged" than any that could be made by merely human artificers, but it was a nonetheless a machine. And as such its workings could be accounted for in just the same way that one accounted for the workings of the ingenious automatons that fascinated early modern aristocratic society: statues with hidden springs, gears, and wheels that moved their limbs and even vocalized (see the cock automaton in fig. 6). There was no problem in conceding that the movements of such automatons were produced by wholly mechanical causes—after all, human beings

7. Strictly speaking (and as noted above), the human body that was to be explained in these mechanical terms was not that of actual, living, breathing human beings but that of beings Descartes *imagined* and offered as conceptual analogies for actual human beings. Here, as in Descartes's overall explanatory framework, what was to be mechanically explained was not the body or world that God *did* create but one he *might* have created. Such an imagined body, or imagined natural world, was judged to be rationally comprehensible and not to conflict significantly with what was reliably known about the workings of the real body and real natural world. And it was this judgment that suggested the possibility that Descartes's account of the imagined body or world might serve for the real one. In fact, many readers set aside the caveat and took Descartes's distinction between the real and the imaginary as merely a culturally expedient rhetorical device.

constructed them—and there was, for Descartes, also no problem in construing every aspect of *animal* bodies in similarly mechanical terms. There was nothing about monkeys, or about honeybees, that could not, in Descartes's view, be adequately accounted for by their physical mechanisms.

For human beings, however, the scope of mechanical accounts was crucially limited. Explanations of the human *body* were, for Descartes, not the same thing as explanations of human *beings,* for there was something about human beings that could not be comprehended by an account of the body's matter and motion. We do not *feel* ourselves to be machines, and Descartes agreed that we are not. We feel ourselves to exercise will, to have purposes, to move our bodies in response to our purposes, to be conscious, to make moral evaluations, to deliberate and to reason (that is, to *think*), and to express the results of our thought in language—none of which Descartes reckoned that machines, or animals, can do.[8]

That human beings have these attributes and can do these things arises from their *dual* nature: as far as their bodies are concerned, they are matter in motion, but they also have *minds,* and the phenomena of mind are not ultimately to be accounted for by matter in motion. The world itself contains two qualitatively different realms, that of matter and that of mind. It is only in human beings that the two realms meet. Human beings, alone of God's animate creations, possess a "rational soul." This soul is the special endowment of God; it is what links human beings to their Creator; and it is what links Descartes's philosophical account to Scripture. Every human soul is specially created by God; it is immortal; and unlike matter, it is neither extended in space nor divisible. This most ambitious of seventeenth-century programs of mechanical explanation also ended in mystery.

The mystery concerns *how* mind and matter meet in the human frame. There were analogies available for speaking of such myste-

8. By no means all early modern thinkers found such a radical distinction between human and animal capacities persuasive: the late sixteenth-century essayist Michel de Montaigne, for example, was disposed to attribute to animals feelings, reason, and even genuine language.

rious unions—the union of a rock with gravity, of the fingers with the hand, of different kinds of tissue in the same body—but in the end what united mind and matter in human beings was a primary, and therefore unexplicated, notion. If the mind is unextended in space, *where* is it? At what place do the two realms make contact? Here Descartes did offer a candidate response. Just as all sensations and impressions must come together to be the objects of thought, so one is to look for a little organ, not duplicated in bilateral symmetry, in the middle of the mass of the brain (fig. 30). This was the small pineal gland, "the seat of imagination and common sense," indeed, "the seat of the soul." Tiny, and tenuously supported only by surrounding blood vessels, it was well adapted to transfer movements from the body to the mind and from the mind to the body. The ultimate mystery resided, fittingly, in a *point.*

Pressing his program of mechanical explanation as far as he could make it go, Descartes ended with a notion that was itself outside the scope of his mechanical philosophy and that even appeared to violate some of his most cherished principles. The uniqueness of human beings flowed from the mysterious interaction between what could be encompassed within a mechanical framework and what could not. Human beings have purposive minds, and purposive minds, after all, move matter. As you turn the pages of this book, you manifest the causal role of mind in nature. And just as mechanism was limited by both religious sensibilities and the lived experience of being human, so the rejection of anthropocentrism (described in chapter 1) was limited by the *unintelligibility* of a wholly mechanistic account of what it is to be human. Other natural philosophers, as we have seen, set limits on mechanical explanations more cautiously than Descartes did, and his accounts were widely suspect for their allegedly subversive effects on religiously approved appreciations of human beings' spiritual nature and unique relationship with God. Yet Descartes's ambitious mechanical program did not deny the special and central place of human beings in the natural world, but offered another idiom for appreciating it. The point at which the mystery of human nature resided was the point at which anthropo-

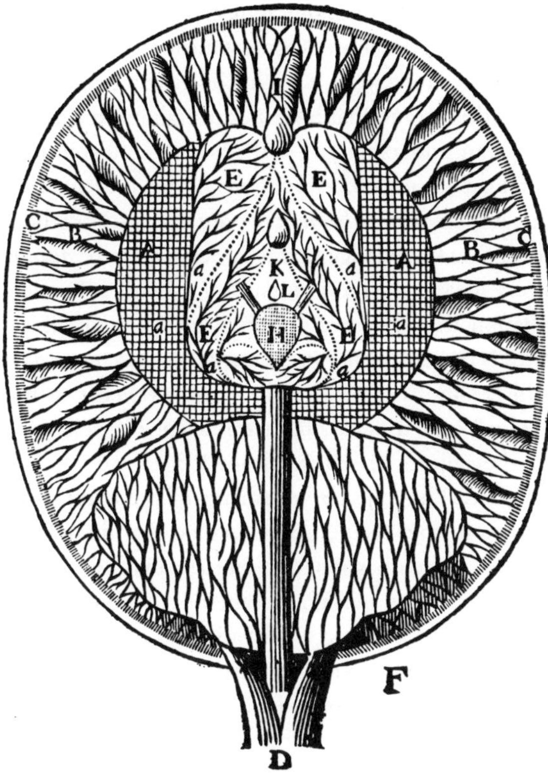

30. *Section of the human brain as depicted by Descartes. The pineal gland is at H. Source: René Descartes,* Treatise of Man *(1664).*

centrism persisted. The special place of human beings in nature was at once a solution to a problem and a problem that might come to demand solution, a legacy of the Scientific Revolution to its cultural heirs.

Disinterestedness and the Uses of Natural Knowledge

I have said that there is nothing like an "essence" of the Scientific Revolution, and I have sought wherever possible to introduce readers to the heterogeneity, and even the contested status, of natural knowledge in the seventeenth century. I do not mean to go back on that self-denying ordinance now, yet I want in conclusion to draw special attention to another strand that has run throughout this narrative and that connects our understanding of the Scientific Revolution to some fundamental categories and evaluations of our present-day culture. That theme is the *depersonalization* of nature and the attendant practices of producing knowledge that is understood to be *disinterested.*

The very idea of the modern natural sciences is bound up with an appreciation that they are objective rather than subjective accounts. They represent *what is* in the natural world, not *what ought to be,* while the possibility of such a radical distinction between scientific "is-knowledge" and moral "ought-knowledge" itself depends on separating the objects of natural knowledge from the objects of moral discourse. The objective character of the natural sciences is supposed to be further secured by a method that disciplines practitioners to set aside their passions and interests in the making of scientific knowledge. Science, in this account, fails to report objectively on the world—it fails to *be* science—if it allows considerations of value, morality, or politics to intrude into the processes of making and validating knowledge. When science is being done, society is kept at bay. The broad form of this understanding of science was developed in the seventeenth century, and that is one major reason canonical accounts have identified the Scientific Revolution as the epoch that made the world modern.

It is difficult even to describe these achievements without being heard to endorse them: How else could properly scientific knowledge be produced? In what other ways could we have secure natural knowledge? The historian's task, of course, is neither to praise nor to blame. For all that, it is worth noting that the modern boundaries that have sought to place the explicitly subjective and moral on the other side of the properly scientific have had interesting consequences in our culture. One effect has been to deny that there can be such a thing as a science of values. Talk of moral good and bad is understood to be arbitrary, interested, and irresolvable by reason, whereas talk of what exists in the natural world can be rational, disinterested, and consensual. That sentiment too was an achievement of the Scientific Revolution and its immediate legacy. And that too is closely identified with the modern condition.

The Aristotelian teleological framework so vigorously criticized by the mechanical philosophers offered an integrated understanding of the human and the natural, with a teleological idiom deemed proper for interpreting both. But seventeenth-century mechanical philosophers' rejection of teleology meant that ways of talking about human ends were to differ fundamentally from ways of talking about natural processes. Human beings may no longer reside at the center of the cosmos, but modern ways of speaking about our sentient and moral nature have become even more special and more separated from the idiom of "the natural." Human bodies might be machines—as mechanical philosophers like Descartes insisted they were—but a condition of our collective humanity remains the presumption that we treat fellow human beings as if they are not machines. Talk among human beings trades in the notion of purpose, consciousness, and moral accountability. That is how we practically distinguish ourselves from machines and from mechanical nature. This too is central to the modern condition: our success in understanding nature has generated deep problems for understanding our place in it and, indeed, for understanding *human nature.*

Similarly, the reform of natural knowledge in the seventeenth century meant that practitioners achieved relative confidence in their accounts of what the real underlying structure of the natural world was like—corpuscular, mechanical, mathematical—at the price of breaking up a traditional connection between how things appear to us and how we (officially) think they really are. It might therefore be said that the success of natural science, and especially its capacity to generate consensus, has been secured at the cost of separating itself from a practice *now* to be called "philosophy" and in particular the philosophy of *knowledge.* We now know, with great confidence and certainty, about the natural world—science—but our understanding of *how it is* that we can know about that world—modern philosophy—is a notably divisive, disputatious, and, some would even (uncharitably) say, unsatisfactory area of our present-day culture. Good order and certainty in science have been produced at the price of disorder and uncertainty elsewhere in our culture.

Finally, one can draw attention to a great paradox that lies at the heart of modern science and that was, arguably, put there in the seventeenth century. The paradox again concerns the relation between the objective and disinterested identity of the natural sciences and the everyday world of subjectivity, passions, and interests. This is the paradox: the more a body of knowledge is understood to be objective and disinterested, the more valuable it is as a tool in moral and political action. Conversely, the capacity of a body of knowledge to make valuable contributions to moral and political problems flows from an understanding that it was not produced and evaluated to further particular human interests. That paradox is also a legacy of the Scientific Revolution, when disengaged scholars and gentlemen forged a body of knowledge that was enormously useful for theology and politics precisely because its practitioners advertised the boundaries between "science" and "affairs of church and state." So too for late twentieth-century moderns: the most powerful storehouse of value in our modern culture is the body of knowledge we consider to have least to do with the discourse of moral value.

One consequence of the presentation of science developed in the seventeenth century—to be sure, one of the least important—is that many of the categories we have available for talking about science are just those whose history and sociology we wish to understand. So, for example, if we seek to understand "the influence of society on science," or "the relationship between science and values," we run the risk of taking for granted the existence of entities whose distinctiveness in our culture was a product of the Scientific Revolution. It has been suggested that unless we invent a special new language to talk about these things we will remain trapped in an unsatisfactory "modern" condition. I am rather more optimistic. I think that as we come to understand more about the processes that made our culture what it is, words like "science" and "society" will come to have new meanings. And I have some hope that at least a few readers will think differently about science and society now than they did at the beginning of this book.

One final awkwardness remains to be confronted. The cultural inheritance that enjoins us to contrast the domain of science to that of human passions and interests counts not only as description but also as prescription: this is how things *ought* to be arranged in science. That means that any account, such as this one, and much recent history and sociology of science, that seeks to portray science as the contingent, diverse, and at times deeply problematic product of interested, morally concerned, historically situated people is likely to be read as *criticism* of science. It may be thought that anyone making such claims must be motivated by a desire to expose science—to say that science is not objective, not true, not reliable—or that such accounts will have the effect of eroding respect for science.

This, in my view, would be both an unfortunate and an inaccurate conclusion. Something *is* being criticized here: it is not *science* but some pervasive *stories* we tend to be told about science. Most critics of science happen to be scientists, and I think they are far better placed to do that critical job than historians, sociologists, or philosophers. Science remains whatever it is—certainly the most reliable body of natural knowledge we have got—whether the stories we are told about its historical development and social relations are accurate or inaccurate. Science remains also the most respected component of our modern culture. I doubt very much whether science needs to be defended through perpetuating fables and myths cobbled together to pour value over it. To do so would truly be the final denial of the cultural legacy of the Scientific Revolution.

Bibliographic Essay

It is both an obligation and a privilege to acknowledge the extent to which this brief account of the Scientific Revolution is indebted to an immense body of work produced over many years by many other historians. No reader should take this little book as much more than a synthetic sketch of a small portion of that highly detailed scholarship and an invitation to satisfy curiosity by reading further. The following selective bibliographic essay should serve to guide interested readers to much of the relevant historical work (confined almost exclusively to Anglophone and translated texts). An asterisk (*) marks the writings I have relied on most heavily in this account.

1. The "Great Tradition" in the History of Science

Those setting out to acquaint themselves with the identity of the Scientific Revolution, and with its major actors, themes, problems, achievements, and conceptual resources, can draw on a distinguished body of what now is commonly called "traditional" scholarship. If indeed it *is* traditional, that is because this literature typically manifested robust confidence that there was a coherent and specifiable body of early modern culture rightly called revolutionary, that this culture marked a clear break between "old" and "new," that it had an "essence," and that this essence could be captured through accounts of the rise of mechanism and materialism, the mathematization of natural philosophy, the emergence of a full-blooded experimentalism, and for many, though not all, traditional writers, the identification of an effective "method" for producing authentic science.

Among the outstanding achievements of this type of scholarship are the early work of *E. A. Burtt, *The Metaphysical Foundations of Modern Physical Science* (New York: Doubleday Anchor, 1954; orig. publ. 1924); A. C. Crombie, *Augustine to Galileo: The History of Science, A.D. 400–1650* (London: Falcon, 1952); A. Rupert Hall, *The Scientific Revolution, 1500–1800: The Formation of the Modern Scientific Attitude,* 2d ed. (Boston: Beacon Press, 1966; orig. publ. 1954); idem, *From Galileo to Newton, 1630–1720* (London: Collins, 1963); Marie Boas [Hall], *The Scientific Renaissance, 1450–1630* (New York: Harper Torchbooks, 1966; orig. publ. 1962); and E. J. Dijksterhuis, *The Mechanization of the World Picture: Pythagoras to Newton,* trans. C. Dikshoorn (Princeton: Princeton University Press, 1986; orig. publ.

1950). Herbert Butterfield's *The Origins of Modern Science, 1300–1800,* rev. ed. (New York: Free Press, 1965; orig. publ. 1949) is a highly influential account of the Scientific Revolution designed for a general historical readership, as are relevant portions of the general survey of the history of science from the Renaissance onward in Charles C. Gillispie's *The Edge of Objectivity: An Essay in the History of Scientific Ideas* (Princeton: Princeton University Press, 1990; orig. publ. 1960), esp. chaps. 2–4. The polemical contest between reformers and traditionalists in English science was the subject of Richard Foster Jones's classic *Ancients and Moderns: A Study of the Rise of the Scientific Movement in Seventeenth-Century England* (New York: Dover Books, 1981; orig. publ. 1962); see also Joseph M. Levine, "Ancients and Moderns Reconsidered," *Eighteenth Century Studies* 15 (1980–81): 72–89. I. Bernard Cohen has concisely traced the historical development of the very idea of a scientific revolution in *"The Eighteenth-Century Origins of the Concept of Scientific Revolution," *Journal of the History of Ideas* 37 (1976): 257–88, very much expanded in his **Revolution in Science* (Cambridge: Harvard University Press, 1985); see also Amos Funkenstein, "Revolutionaries on Themselves," in *Revolutions in Science: Their Meaning and Relevance,* ed. William R. Shea (Canton, Mass.: Science History Publications, 1988), 157–63. *A. Rupert Hall's essay "On the Historical Singularity of the Scientific Revolution of the Seventeenth Century," in *The Diversity of History: Essays in Honour of Sir Herbert Butterfield,* ed. J. H. Elliott and H. G. Koenigsberger (London: Routledge and Kegan Paul, 1970), 199–222, is a forceful assertion of the uniqueness, coherence, and power of *the* Scientific Revolution against emerging revisionist voices.

Butterfield's work, like that of the Halls, Gillispie, and several other post–World War II historians, shows the strong influence of Alexandre Koyré, whose writings were concerned inter alia to show how the rational physical science of the seventeenth century constituted a radical break with common sense and with the dictates of sensory experience. For that reason, Koyré tended to downplay the significance of the experimentalism and inductivism that characterized much English science in the seventeenth century. For him, Galilean idealization and rationalism were closest to the "essence" of the Scientific Revolution: see especially Koyré's *Galileo Studies,* trans. John Mepham (Atlantic Highlands, N.J.: Humanities Press, 1978; orig. publ. 1939); *From the Closed World to the Infinite Universe* (Baltimore: Johns Hopkins University Press, 1968; orig. publ. 1957); *Newtonian Studies* (London: Chapman and Hall, 1965); and *Metaphysics and Measurement: Essays in Scientific Revolution* (Cambridge: Harvard University Press, 1968);

and for an important French philosophical inspiration to Koyré's rationalism and his stress on the discontinuity of scientific change, see Gaston Bachelard, *The New Scientific Spirit,* trans. Arthur Goldhammer (Boston: Beacon Press, 1984; orig. publ. 1934). Mathematical physics is at its customary center-stage place in concise and accessible surveys by *I. Bernard Cohen, *The Birth of a New Physics,* rev. ed. (New York: W. W. Norton, 1985; orig. publ. 1960), and *Richard S. Westfall, *The Construction of Modern Science: Mechanisms and Mechanics* (Cambridge: Cambridge University Press, 1977; orig. publ. 1971).

I have characterized "traditional" scholarship in this area partly by its adherence to the view that the Scientific Revolution did mark a sharp and definitive break with what went before. Yet throughout the early part of this century and into more recent decades, a minority among eminent historians strove to show significant *continuities*—in concepts and in practices—between medieval science and that of the sixteenth and seventeenth centuries. This was a sensibility especially attractive to scholars well versed in Aristotelian traditions in natural philosophy and therefore less prone to take "modern" polemical critiques of Scholasticism simply as adequate accounts of historical reality. In this connection the early twentieth-century writings of the French physicist and philosopher Pierre Duhem were especially influential: see, for example, *The Aim and Structure of Physical Theory,* trans. Philip P. Wiener (Princeton: Princeton University Press, 1991; orig. publ. 1906), and for more recent texts, A. C. Crombie, *Robert Grosseteste and the Origins of Experimental Science, 1100–1700* (Oxford: Clarendon Press, 1953); Charles B. Schmitt, "Towards a Reassessment of Renaissance Aristotelianism," *History of Science* 11 (1973): 159–93; idem, *Aristotle and the Renaissance* (Cambridge: Harvard University Press, 1983); see also Peter Dear's *Discipline and Experience* (p. 192 below) for rich accounts of the vitality of Aristotelian practice into the seventeenth century.

2. *Historiographical Revisions and Debates*

Traditional views of the Scientific Revolution have been hotly disputed, and even rejected, by some recent historians. Grounds of dissent have varied, but in one way or another this newer work tends to be skeptical of the coherence and integrity of what had previously been understood as the Scientific Revolution. Revisionist historiography is suspicious of talk about its "essence," its coherently and effectively methodical character, and its unambiguous "modernity." This newer scholarship is reluctant to take as its major task celebrating the heroic achievements of "Great Men Making Modernity," preferring to interpret historical figures' aims in "contextual" and often in mundane terms. Such work often seeks to uncover the voices of "lesser" participants (and sometimes of the laity) and to trace the role of forms of culture traditionally considered peripheral to, or even outside, "science proper." Within the past fifteen years or so, some historians—without necessarily rejecting the conceptual identity of the Scientific Revolution—have become intensely interested in the concrete *practices* through which scientific concepts (and even scientific facts) were produced. Debates over the proper description and interpretation of the Scientific Revolution have tended to develop a highly reflective "historiographical" character: what you say about the Scientific Revolution is now widely presumed to implicate fundamental conceptions of what it is, in general, to produce an authentically historical account.

Formal writings on the historiography of the Scientific Revolution have had a partisan quality for a long time, reflecting the historical community's deep-rooted disagreements about what it is that needs to be interpreted and how it is best interpreted. A thorough recent survey of some of these issues is H. Floris Cohen, *The Scientific Revolution: A Historiographical Inquiry* (Chicago: University of Chicago Press, 1994). Its bibliography is a useful starting point, but some of Cohen's characterizations of other historians' views must be treated with caution. A balanced historiographic survey that lucidly sets out many of the issues involved in giving a genuinely historical account of the Scientific Revolution is *Roy Porter's "The Scientific Revolution: A Spoke in the Wheel?" in *Revolution in History,* ed. Roy Porter and Mikuláš Teich (Cambridge: Cambridge University Press, 1986), 290–316, while *Reappraisals of the Scientific Revolution,* ed. David C. Lindberg and Robert S. Westman (Cambridge: Cambridge University Press, 1990), contains a number of excellent papers, of which special note should be taken here of Lindberg's introductory essay—"Conceptions of the Scientific Revolution from Bacon to Butterfield: A Preliminary Sketch," 1–26—and *Ernan McMullin's "Conceptions of Science in the Scientific Revolution," 27–92, which nicely surveys historical variation in the definition of science and in appropriate methodology. For overviews of relevant national differences in science, *The Scientific Revolution in National Context,* ed. Roy Porter and Mikuláš Teich (Cambridge: Cambridge University Press, 1992), has some fine historiographical essays (and see the substantial essay-review of this book by Lorraine Daston, "The Several Contexts of the Scientific Revolution," *Minerva* 32 [1994]: 108–14).

Although in traditional accounts fundamental changes in mathematical physics constitute the "essence" of the Scientific Revolution, the culture of physics and the mathematical sciences is not coextensive with "early modern science." A seminal essay distinguishing between traditions of scientific practice—only some of which are said to have been "revolutionized" during the seventeenth century—is *Thomas S. Kuhn's "Mathematical versus Experimental Traditions in the Development of Physical Science," in his *The Essential Tension: Selected Studies in Scientific Tradition and Change* (Chicago: University of Chicago Press, 1977), 31–65. Kuhn's celebrated *The Structure of Scientific Revolutions,* 2d ed. (Chicago: University of Chicago Press, 1970; orig. publ. 1962) offers an influential general framework for appreciating the nature of revolutionary change in science, while at the same time contributing a pluralist sensibility toward the range of practices that might collectively be called "scientific."

Some of the most vigorous skepticism about the coherence and identity of the Scientific Revolution has been expressed by John A. Schuster in "The Scientific Revolution," in *Companion to the History of Modern Science,* ed. R. C. Olby et al. (London: Routledge, 1990), 217–42. Schuster's skepticism denies the existence of a single, coherent, and efficacious *method* for science, an argument he has made most forcefully in a series of essays on Descartes, the seventeenth-century philosopher who claimed most for the power of method: Schuster, "Cartesian Method as Mythic Speech: A Diachronic and Structural Analysis," in *The Politics and Rhetoric of Scientific Method: Historical Studies,* ed. John A. Schuster and Richard R. Yeo (Dordrecht: D. Reidel, 1986), 33–95; idem, "Whatever Should We Do with Cartesian Method?—Reclaiming Descartes for the History of Science," in *Essays on the Philosophy and Science of René Descartes,* ed. Stephen Voss (New York: Oxford University Press, 1993), 195–223. For well-argued skepticism about the defining place of metaphysics, mechanism, and mathematics in the Scientific Revolution, see Wilson, *The Invisible World* (p. 194 below), esp. chap. 1, and for reflection on literary-generic aspects of traditional accounts of the Scientific Revolution, see Rivka Feldhay, "Narrative Constraints on Historical Writing: The Case of the Scientific Revolution," *Science in Context* 7 (1994): 7–24.

During the period extending roughly from the Second World War to the end of the Cold War, the historiography of the Scientific Revolution was strongly shaped by a general debate over the respective role of "internal" versus "external" factors (as usage then pervasively had it). Internalist historians held that the development of science was adequately accounted for by pointing to the role of evidence, reason, and method, and they liked to refer to the sufficient internal or "immanent" logic of scientific development. Externalists tended to argue that the causal role of these "intellectual factors" had to be supplemented by that of factors deemed external to science, such as the political, religious, or economic concerns of the social and cultural context in which science was being shaped. The argument between internalists and externalists—now largely given up—was not particularly coherent or well focused, but it was highly charged with ideological meaning. Externalists often tended to be Marxists or sympathetic to Marxism, while self-conscious internalism developed partly as a response from scholars who saw the causal invocation of external "social factors" as a form of denigration or even as the "vulgar Marxist" arm of the communist threat to the Free World.

The curious but consequential history of these historiographical debates is briefly reviewed in Steven Shapin, "Discipline and Bounding: The History and Sociology of Science as Seen through the Externalism-Internalism Debate," *History of Science* 30 (1992): 333–69. The classic bête noir of Marxist externalism is a long essay by the Soviet physicist and philosopher Boris Hessen, "The Social and Economic Roots of Newton's 'Principia,'" in *Science at the Cross Roads,* ed. N. I. Bukharin et al. (London: Frank Cass, 1971; orig. publ. 1931), 149–212; see also Edgar Zilsel, "The Sociological Roots of Science," *American Journal of Sociology* 47 (1942): 245–79; Franz Borkenau, "The Sociology of the Mechanistic World-Picture," *Science in Context* 1 (1987): 109–27 (art. orig. publ. 1932); Henryk Grossmann, "The Social Foundations of Mechanistic Philosophy and Manufacture," *Science in Context* 1 (1987): 129–80 (art. orig. publ. 1935); also see George Clark, *Science and Social Welfare in the Age of Newton,* 2d ed. (Oxford: Clarendon Press, 1970; orig. publ. 1937); and Robert M. Young, "Marxism and the History of Science," in *Companion to the History of Modern Science* (p. 172 above), 77–86. The Marxist tradition in the study of science has recently lost much of its former vigor, but it is by no means defunct: for recent continuations of broadly Marxist sensibilities in the study of early modern science, see works by James Jacob and Margaret Jacob (pp. 204–5 below); Richard Hadden and Frank Swetz (both p. 180 below); and Gideon Freudenthal, *Atom and Individual in the Age of Newton: On the Genesis of the Mechanistic World View,* trans. Peter McLaughlin (Dordrecht: D. Reidel, 1986; orig. publ. 1982). The 1938 work of the American sociologist Robert K. Merton on the relation between science and religion in seventeenth-century England (p. 195 below), though it carefully dissociated itself from Marxist externalism, was nevertheless also an important target of attack, and the historiography of Alexandre Koyré (p. 169 above and p. 177 below) was an important resource in the internalist response to both Marxism and Merton.

Related historiographical issues of the social identity of the scientific practitioner and the social relations of science were canvassed in A. Rupert Hall, "The Scholar and the Craftsman in the Scientific Revolution," in *Critical Problems in the History of Science,* ed. Marshall Clagett (Madison: University of Wisconsin Press, 1959), 3–23, and they continue to occupy historians of the Scientific Revolution: see, for example, contributions to *Renaissance and Revolution: Humanists, Scholars, Craftsmen and Natural Philosophers in Early Modern Europe,* ed. J. V. Field and Frank A. J. L. James (Cambridge: Cambridge University Press, 1993), and *Steven Shapin, "'A Scholar and a Gentleman': The Problematic Identity of the Scientific Practitioner in Early Modern England," *History of Science* 29 (1991): 279–327.

3. Frameworks and Disciplines

A. The Mechanical Philosophy and the Physical Sciences

Mechanism and related issues are centrally treated in all traditional accounts of the Scientific Revolution—in fact, for most standard treatments they effectively constitute that revolution. In addition to works cited in section 1 above, see especially *Marie Boas [Hall], "The Establishment of the Mechanical Philosophy," *Osiris* 10 (1952): 412–541 (also a good starting source for matter theory). *J. A. Bennett, "The Mechanics' Philosophy and the Mechanical Philosophy," *History of Science* 24 (1986): 1–28, is a superb essay on the relation between mechanical thinking about nature and the role of mechanics, while *Otto Mayr's *Authority, Liberty and Automatic Machinery in Early Modern Europe* (Baltimore: Johns Hopkins University Press, 1986) is the best source for the general cultural significance of clockwork and the clock metaphor; see also Klaus Maurice and Otto Mayr, eds., *The Clockwork Universe: German Clocks and Automata, 1550–1650* (New York: Neale Watson, 1980); Derek J. de Solla Price, "Automata and the Origins of Mechanism and Mechanistic Philosophy," *Technology and Culture* 5 (1964): 9–23; Silvio A. Bedini, "The Role of Automata in the History of Technology," *Technology and Culture* 5 (1964): 24–42; and *Laurens Laudan, "The Clock Metaphor and Probabilism: The Impact of Descartes on English Methodological Thought, 1650–65," *Annals of Science* 22 (1966): 73–104. For Mersenne and mechanism, see Robert Lenoble, *Mersenne, ou La naissance du* *mécanisme* (Paris: J. Vrin, 1943); Peter Dear, *Mersenne and the Learning of the Schools* (Ithaca: Cornell University Press, 1988), chap. 6; and Gaukroger's biography of Descartes (p. 209 below), 146–52 (and chap. 3 for Isaac Beeckman). Fundamental problems with defining the coherence, intelligibility, and cultural identity of the mechanical philosophy are treated in *Alan Gabbey's important paper "The Mechanical Philosophy and Its Problems: Mechanical Explanations, Impenetrability, and Perpetual Motion," in *Change and Progress in Modern Science,* ed. Joseph C. Pitt (Dordrecht: D. Reidel, 1985), 9–84; see also idem, "The Case of Mechanics: One Revolution or Many?" in *Reappraisals of the Scientific Revolution* (p. 171 above), and *Alan Chalmers, "The Lack of Excellency of Boyle's Mechanical Philosophy," *Studies in History and Philosophy of Science* 24 (1993): 541–64. And for problems with the intelligibility of Newton's treatment of gravitation, see *Gerd Buchdahl, "Gravity and Intelligibility: Newton to Kant," in *The Methodological Heritage of Newton,* ed. Robert E. Butts and John W. Davis (Toronto: University of Toronto Press, 1970), 74–102. For acute historical sensibilities toward the identity of the practice known as "natural philosophy," with special reference to its Newtonian form, see Simon Schaffer, "Natural Philosophy," in *The Ferment of Knowledge: Studies in the Historiography of Eighteenth-Century Science,* ed. George S. Rousseau and Roy Porter (Cambridge: Cambridge University Press, 1980), 55–91.

The range of physical sciences is of course extensively discussed in all of the "Great Tradition" works noted here and in section 1 above. For additional topics in physics, see Richard S. Westfall, *Force in Newton's Physics: The Science of Dynamics in the Seventeenth Century* (London: Macdonald, 1971); I. Bernard Cohen, *Franklin and Newton: An Inquiry into Speculative Newtonian Experimental Science and Franklin's Work in Electricity as an Example Thereof* (Cambridge: Harvard University Press, 1966; orig. publ. 1956), esp. chaps. 5–6; idem, *The Newtonian Revolution* (Cambridge: Cambridge University Press, 1980); John Heilbron, *Electricity in the Seventeenth and Eighteenth Centuries: A Study of Early Modern Physics* (Berkeley: University of California Press, 1979), and the abridgement in idem, *Elements of Early Modern Physics* (Berkeley: University of California Press, 1982); Mary B. Hesse, *Forces and Fields: A Study of Action at a Distance in the History of Physics* (Totowa, N.J.: Littlefield, Adams, 1965; orig. publ. 1961); A. I. Sabra, *Theories of Light: From Descartes to Newton* (Cambridge: Cambridge University Press, 1981; orig. publ. 1967); Alan E. Shapiro, *Fits, Passions, and Paroxysms: Physics, Method, and Chemistry and Newton's Theories of Colored Bodies and Fits of Easy Reflection* (Cambridge: Cambridge University Press,

1993), pt. 1; idem, "Kinematic Optics: A Study of the Wave Theory of Light in the Seventeenth Century," *Archive for History of Exact Sciences* 11 (1973): 134–266; and Edward Grant, *Much Ado about Nothing: Theories of Space and Vacuum from the Middle Ages to the Scientific Revolution* (Cambridge: Cambridge University Press, 1981). For classic sources on the Torricellian experiment and related issues in pneumatics, see Cornélis de Waard, *L'expérience barométrique: Ses antécédents et ses explications* (Thouars: J. Gamon, 1936), and Jean-Pierre Fanton d'Andon, *L'horreur du vide: Expérience et raison dans la physique pascalienne* (Paris: CNRS, 1978); for a recent intellectual biography of Pascal, see Donald Adamson, *Blaise Pascal: Mathematician, Physicist, and Thinker about God* (New York: St. Martin's Press, 1995); and for aspects of magnetism, Stephen Pumfrey, "Mechanizing Magnetism in Restoration England: The Decline of Magnetic Philosophy," *Annals of Science* 44 (1987): 1–22, and idem, "'O tempora, O magnes!' A Sociological Analysis of the Discovery of Secular Magnetic Variation in 1634," *British Journal for the History of Science* 22 (1989): 181–214.

B. General Views of Nature and the Environment

A number of classic texts trace the broad outlines of changing views of nature from the Renaissance through the Scientific Revolution, usually drawing special attention to the shift from organicist to mechanical conceptions. An excellent starting point is R. G. Collingwood's *The Idea of Nature* (London: Oxford University Press, 1960; orig. publ. 1945), esp. pt. 2, chap. 1, and with reference to the "microcosm/macrocosm" scheme and the hierarchical interconnectedness of creation, see one of the defining exercises in the history of metaphysical ideas, Arthur O. Lovejoy, *The Great Chain of Being: A Study of the History of an Idea* (Cambridge: Harvard University Press, 1964; orig. publ. 1936), and also E. M. W. Tillyard, *The Elizabethan World Picture* (Harmondsworth: Pelican, 1972; orig. publ. 1943). For "physicotheology" and notions of "the environment," see Clarence J. Glacken, *Traces on the Rhodian Shore: Nature and Culture in Western Thought from Ancient Times to the End of the Eighteenth Century* (Berkeley: University of California Press, 1976; orig. publ. 1967), pt. 3; Yi-fu Tuan, *The Hydrologic Cycle and the Wisdom of God: A Theme in Geoteleology* (Toronto: University of Toronto Press, 1968); idem, *Topophilia: A Study of Environmental Perception, Attitudes, and Values* (Englewood Cliffs, N.J.: Prentice-Hall, 1974); and Roy Porter, "The Terraqueous Globe," in *The Ferment of Knowledge* (p. 175

above), 285–324. For more social-historically oriented approaches to changing views of nature, see Keith Thomas, *Religion and the Decline of Magic: Studies in Popular Beliefs in Sixteenth- and Seventeenth-Century England* (Harmondsworth: Penguin, 1973), and idem, *Man and the Natural World: A History of the Modern Sensibility* (New York: Pantheon, 1983). See also Allen G. Debus, *Man and Nature in the Renaissance* (Cambridge: Cambridge University Press, 1978), and for feminist perspectives see, for example, Carolyn Merchant, *The Death of Nature: Women, Ecology and the Scientific Revolution* (San Francisco: Harper San Francisco, 1990; orig. publ. 1980).

C. Astronomy and Astronomers

Copernicanism and related issues in theoretical and observational astronomy have been thoroughly discussed in "Great Tradition" texts on the Scientific Revolution. A useful and concise entry to this literature is J. R. Ravetz, "The Copernican Revolution," in *Companion to the History of Modern Science* (p. 172 above), 201–16, while a detailed account of technical and conceptual issues is Thomas S. Kuhn, *The Copernican Revolution: Planetary Astronomy in the Development of Western Thought* (Cambridge: Harvard University Press, 1957). Arthur Koestler's semipopular treatment of Kepler, Tycho Brahe, and Galileo, *The Sleepwalkers: A History of Man's Changing Vision of the Universe* (New York: Macmillan, 1959), still has the capacity to stimulate and provoke. See also Alexandre Koyré, *The Astronomical Revolution: Copernicus—Kepler—Borelli,* trans. R. E. W. Maddison (New York: Dover Books, 1992; orig. publ. 1973), and his *From the Closed World to the Infinite Universe* (p. 169 above); Albert Van Helden, *Measuring the Universe: Cosmic Dimensions from Aristarchus to Halley* (Chicago: University of Chicago Press, 1985); Karl Hufbauer, *Exploring the Sun: Solar Science since Galileo* (Baltimore: Johns Hopkins University Press, 1991), 1–32; Edward Grant, *Planets, Stars, and Orbs: The Medieval Cosmos, 1200–1687* (Cambridge: Cambridge University Press, 1994); Jean Dietz Moss, *Novelties in the Heavens: Rhetoric and Science in the Copernican Controversy* (Chicago: University of Chicago Press, 1994); James M. Lattis, *Between Copernicus and Galileo: Christoph Clavius and the Collapse of Ptolemaic Cosmology* (Chicago: University of Chicago Press, 1994); various essays in *The Copernican Achievement,* ed. Robert S. Westman (Berkeley: University of California Press, 1975); Westman, "The Copernicans and the Churches," in *God and Nature: Historical Essays on the Encounter between Christianity and Science,* ed. David C. Lindberg and Ronald L.

Numbers (Berkeley: University of California Press, 1986), 76–113; and Westman's influential assessment of the early modern astronomer's disciplinary identity, "The Astronomer's Role in the Sixteenth Century: A Preliminary Study," *History of Science* 18 (1980): 105–47.

An excellent account of the career of Tycho Brahe is Victor E. Thoren, *The Lord of Uraniborg: A Biography of Tycho Brahe* (Cambridge: Cambridge University Press, 1990); for Kepler see, for example, Max Caspar, *Kepler, 1571–1630,* ed. and trans. C. Doris Hellman (New York: Collier Books, 1962; orig. publ. 1959); Nicholas Jardine, *The Birth of History and Philosophy of Science: Kepler's "A Defence of Tycho against Ursus" with Essays on Its Provenance and Significance* (Cambridge: Cambridge University Press, 1984); Bruce Stephenson, *Kepler's Physical Astronomy* (Princeton: Princeton University Press, 1994; orig. publ. 1987); and idem, *The Music of the Heavens: Kepler's Harmonic Astronomy* (Princeton: Princeton University Press, 1994), while a series of marvelous recent papers on observational astronomy raise issues of fundamental importance to the understanding of the Scientific Revolution: for example, *Albert Van Helden, "Telescopes and Authority from Galileo to Cassini," *Osiris* 9 (1994): 8–29; Mary G. Winkler and Albert Van Helden, "Representing the Heavens: Galileo and Visual Astronomy," *Isis* 83 (1992): 195–217; idem, "Johannes Hevelius and the Visual Language of Astronomy," in *Renaissance and Revolution* (p. 174 above), 95–114. For treatment of observational and theoretical issues in planetary astronomy, see *Planetary Astronomy from the Renaissance to the Rise of Astrophysics. Part A: Tycho Brahe to Newton,* General History of Astronomy, vol. 2, ed. René Taton and Curtis Wilson (Cambridge: Cambridge University Press, 1989); for cometary astronomy, see James A. Ruffner, "The Curved and the Straight: Cometary Theory from Kepler to Hevelius," *Journal of the History of Astronomy* 2 (1971): 178–94; and for an accessible account of various aspects of astronomical practice, see Lesley Murdin, *Under Newton's Shadow: Astronomical Practices in the Seventeenth Century* (Bristol: Adam Hilger, 1985). Patronage patterns in relation to Galileo's observational astronomy are treated by *Mario Biagioli, *Galileo, Courtier: The Practice of Science in the Culture of Absolutism* (Chicago: University of Chicago Press, 1993), esp. chaps. 1–2, and Richard S. Westfall, "Science and Patronage: Galileo and the Telescope," *Isis* 76 (1985): 11–30. Patronage in connection with Copernicus's work is discussed by Robert S. Westman, "Proof, Poetics, and Patronage: Copernicus's Preface to *De Revolutionibus,*" in *Reappraisals of the Scientific Revolution* (p. 171 above), 167–205. The relation between astronomy and astrology is dealt with in several sources listed above, but see also

Patrick Curry, *Prophecy and Power: Astrology in Early Modern England* (Princeton: Princeton University Press, 1989); Ann Geneva, *Astrology and the Seventeenth-Century Mind: William Lilly and the Language of the Stars* (Manchester: Manchester University Press, 1994); and various essays in *Astrology, Science and Society: Historical Essays,* ed. Patrick Curry (Woodbridge: Boydell and Brewer, 1987), especially Simon Schaffer, "Newton's Comets and the Transformation of Astrology," 219–43. For sociological sensibilities toward astrology and its scientific opposition, see Peter W. G. Wright, "Astrology and Science in Seventeenth Century England," *Social Studies of Science* 5 (1975): 399–422, and idem, "A Study in the Legitimisation of Knowledge: The 'Success' of Medicine and the 'Failure' of Astrology," in *On the Margins of Science: The Social Construction of Rejected Knowledge,* ed. Roy Wallis, Sociological Review Monograph 27 (Keele: University of Keele Press, 1979), 85–102.

D. Mathematics and Mathematicians

For all the traditional stress on the mathematization of natural philosophy as the "essence" of the Scientific Revolution, and for all the wealth of heroic scholarship on the mathematical papers of key figures of early modern science, the historiography of early modern mathematics remains relatively undeveloped in comparison with other strands of scientific practice. Canonical surveys, including accounts of important aspects of early modern mathematics, are Carl B. Boyer, *The History of Calculus and Its Conceptual Development* (New York: Dover Books, 1959; orig. publ. 1949), esp. chaps. 4–5, and J. F. Scott, *A History of Mathematics: From Antiquity to the Beginning of the Nineteenth Century* (London: Taylor and Francis, 1958), esp. chaps. 6–12. An overview of seventeenth-century developments is D. T. Whiteside, "Patterns of Mathematical Thought in the Seventeenth Century," *Archive for History of Exact Sciences* 1 (1961): 179–388, and an important assessment of mathematics in relation to the "intelligibility" of the new mechanics is Michael S. Mahoney, "Infinitesimals and Transcendent Relations: The Mathematics of Motion in the Late Seventeenth Century," in *Reappraisals of the Scientific Revolution* (see p. 171 above), 461–91. For an account of early modern mathematics in relation to contemporary philosophy of mathematics, see Paolo Mancosu, *Philosophy of Mathematics and Mathematical Practices in the Seventeenth Century* (Oxford: Oxford University Press, 1995). Gaukroger's intellectual biography of Descartes (p. 209 below) is particularly rich in material on mathematics and mechanism, as is Westfall's biography of Newton (p. 210 below). For Hobbes's highly controversial mathematical views, see, for example, Douglas M. Jesseph, "Hobbes and Mathematical Method," *Perspectives on Science* 1 (1993): 306–41; William Sacksteder, "Hobbes: The Art of the Geometricians," *Journal of the History of Philosophy* 18 (1980): 131–46; idem, "Hobbes: Geometrical Objects," *Philosophy of Science* 48 (1981): 573–90; and Helena M. Pycior, "Mathematics and Philosophy: Wallis, Hobbes, Barrow, and Berkeley," *Journal of the History of Ideas* 48 (1987): 265–86. Political issues bearing on the dispute between Newton and Leibniz over priority in the invention of the calculus are discussed in A. Rupert Hall, *Philosophers at War: The Quarrel between Newton and Leibniz* (Cambridge: Cambridge University Press, 1980).

Recent work treats developments in arithmetic and practical mathematics in relation to the socioeconomic context, for example, the Marxist or social historical perspectives of Richard W. Hadden, *On the Shoulders of Merchants: Exchange and the Mathematical Conception of Nature in Early Modern Europe* (Albany: State University of New York Press, 1994); Frank Swetz, *Capitalism and Arithmetic: The New Math of the Fifteenth Century* (La Salle, Ill.: Open Court, 1987); and Witold Kula, *Measures and Men,* trans. R. Szreter (Princeton: Princeton University Press, 1986; orig. publ. 1970). Among other recent studies of practical mathematics, see, for example, A. J. Turner, "Mathematical Instruments and the Education of Gentlemen," *Annals of Science* 30 (1973): 51–88; Stephen Johnston, "Mathematical Practitioners and Instruments in Elizabethan England," *Annals of Science* 48 (1991): 319–44; Frances Willmoth, *Sir Jonas Moore: Practical Mathematics and Restoration Science* (Woodbridge: Boydell Press, 1993); J. A. Bennett, "The Challenge of Practical Mathematics," in *Science, Culture and Popular Belief in Renaissance Europe,* ed. Stephen Pumfrey, Paolo L. Rossi, and Maurice Slawinski (Manchester: Manchester University Press, 1991), 176–90; and Mordechai Feingold, *The Mathematicians' Apprenticeship: Science, Universities and Society in England, 1560–1640* (Cambridge: Cambridge University Press, 1984). Mario Biagioli, "The Social Status of Italian Mathematicians, 1450–1600," *History of Science* 27 (1989): 41–95, is interesting for its discussion of the social standing of mathematicians in relation to philosophers, as is his *Galileo, Courtier,* Westman's "The Astronomer's Role" (both p. 178 above), and Dear's *Discipline and Experience* (p. 192 below).

The important relation between probability theory and strands of experimental philosophy is treated in Ian Hacking, *The Emergence of Probability: A Philosophical Study of Early Ideas about Probability, Induction and*

Statistical Inference (Cambridge: Cambridge University Press, 1975); see also Lorraine J. Daston, *Classical Probability in the Enlightenment* (Princeton: Princeton University Press, 1988), chap. 1, and for social statistics, Peter Buck, "Seventeenth-Century Political Arithmetic: Civil Strife and Vital Statistics," *Isis* 68 (1977): 67–84. Loosely related studies of the origins and senses of the notion of "scientific laws" or "laws of nature" include John R. Milton, "The Origin and Development of the Concept of the 'Laws of Nature,'" *Archives Européennes de Sociologie* 22 (1981): 173–95; Jane E. Ruby, "The Origins of Scientific 'Law,'" *Journal of the History of Ideas* 47 (1986): 341–59; Joseph Needham, "Human Laws and the Laws of Nature," *Journal of the History of Ideas* 12 (1951): 3–32; Edgar Zilsel, "Physics and the Problem of Historico-sociological Laws," *Philosophy of Science* 8 (1941): 567–79; and idem, "The Genesis of the Concept of Scientific Law," *Philosophical Review* 51 (1942): 245–67.

E. Chemistry, Alchemy, and Matter Theory

The move from the "pseudoscience" of alchemy to a "proper" science of matter set within a corpuscular, mechanical, and experimental framework is a theme extensively treated in traditional accounts of the Scientific Revolution; see, among many examples, J. R. Partington, *A History of Chemistry*, 4 vols. (London: Macmillan, 1961–70); Marie Boas [Hall], *Robert Boyle and Seventeenth-Century Chemistry* (Cambridge: Cambridge University Press, 1958); Robert P. Multhauf, *The Origins of Chemistry* (New York: F. Watts, 1967); Henry M. Leicester, *The Historical Background of Chemistry* (New York: John Wiley, 1965; orig. publ. 1956), chaps. 9–12; Maurice Crosland, *Historical Studies in the Language of Chemistry* (London: Heinemann, 1962), esp. pts. 1–2; and Hélène Metzger's influential *Les doctrines chimiques en France du début du XVIIe à la fin du XVIIIe siècle* (Paris: Presses Universitaires de France, 1969; orig. publ. 1923). At the same time, some historians writing in this idiom have been unsure whether the achievements of seventeenth-century chemistry really entitle it to "revolutionary" status, and they have referred to a "postponed chemical revolution" dating from the late eighteenth- and early nineteenth-century work of Lavoisier and Dalton (e.g., Butterfield, *The Origins of Modern Science* [p. 169 above], chap. 11).

The tendency in more recent historical writing has been to adopt a less "triumphalist" approach to seventeenth-century changes in chemical thought and practice and to see less clear-cut divisions between alchemy and chemistry. The changing historiography of seventeenth-century chemistry is well reviewed by J. V. Golinski, "Chemistry in the Scientific Revolution," in *Reappraisals of the Scientific Revolution* (p. 171 above), 367–96. Some of the landmarks of the newer tendency include Owen Hannaway, *The Chemists and the Word: The Didactic Origins of Chemistry* (Baltimore: Johns Hopkins University Press, 1975); Charles Webster, *The Great Instauration: Science, Medicine, and Reform, 1626–1660* (London: Duckworth, 1975); Bruce T. Moran, *The Alchemical World of the German Court: Occult Philosophy and Chemical Medicine in the Circle of Moritz of Hessen (1572–1632)* (Stuttgart: Franz Steiner, 1991); Pamela H. Smith, *The Business of Alchemy: Science and Culture in the Holy Roman Empire* (Princeton: Princeton University Press, 1994); and Piyo Rattansi and Antonio Clericuzio, eds., *Alchemy and Chemistry in the Sixteenth and Seventeenth Centuries* (Dordrecht: Kluwer, 1994). And for further material on alchemy and Paracelsianism and their relation to medicine and science, see Walter Pagel, *Paracelsus: An Introduction to Philosophical Medicine in the Era of the Renaissance,* 2d ed. (Basel: S. Karger, 1982; orig. publ. 1958); idem, *Joan Baptista Van Helmont: Reformer of Science and Medicine* (Cambridge: Cambridge University Press, 1982); Allen G. Debus, *The Chemical Philosophy: Paracelsian Science and Medicine in the Sixteenth and Seventeenth Centuries,* 2 vols. (New York: Science History Publications, 1977); idem, *The English Paracelsians* (London: Oldbourne, 1965); Betty Jo Teeter Dobbs, *The Foundations of Newton's Alchemy, or "The Hunting of the Greene Lyon"* (Cambridge: Cambridge University Press, 1975); idem, *The Janus Face of Genius: The Role of Alchemy in Newton's Thought* (Cambridge: Cambridge University Press, 1991); and William R. Newman, *Gehennical Fire: The Lives of George Starkey, an American Alchemist* (Cambridge: Harvard University Press, 1994).

The themes of matter theory, atomism, and corpuscularianism in the Scientific Revolution, and especially the mechanical insistence on an inanimate conception of matter, have also been extensively discussed: for atomism see, among many examples, Robert H. Kargon, *Atomism in England from Hariot to Newton* (Oxford: Clarendon Press, 1966), the concise review in Martin Tamny, "Atomism and the Mechanical Philosophy," in *Companion to the History of Modern Science* (p. 172 above), 597–609, and various contributions to Ernan McMullin, ed., *The Concept of Matter in Modern Philosophy* (Notre Dame: Notre Dame University Press, 1978; orig. publ. 1963), pt.1. For Bacon's matter theory and cosmology, see Graham Rees, "Francis Bacon's Semi-Paracelsian Cosmology," *Ambix* 22 (1975): 81–101; idem, "Francis Bacon's Semi-Paracelsian Cosmology and the Great Instauration,"

Ambix 22 (1975): 161–73; and idem, *Francis Bacon's Natural Philosophy: A New Source* (Chalfont St. Giles: British Society for the History of Science, 1984). For Newtonian conceptions and their influence on chemical thought, see Arnold Thackray, *Atoms and Powers: An Essay on Newtonian Matter-Theory and the Development of Chemistry* (Cambridge: Harvard University Press, 1970), chap. 2; also J. E. McGuire, "Force, Active Principles, and Newton's Invisible Realm," *Ambix* 15 (1968): 154–208; Ernan McMullin, *Newton on Matter and Activity* (Notre Dame: Notre Dame University Press, 1978); and for the involvement of Newtonian matter theory in political conflict, see Steven Shapin, "Of Gods and Kings: Natural Philosophy and Politics in the Leibniz-Clarke Disputes," *Isis* 72 (1981): 187–215, and Margaret Jacob's *The Newtonians* (p. 205 below). For Cartesian conceptions see, for example, Gaukroger's intellectual biography of Descartes (p. 209 below), esp. chaps. 5 and 7. For Boyle, see Marie Boas [Hall], *Robert Boyle* (p. 181 above); various sources in section 5E below; Thomas S. Kuhn, "Robert Boyle and Structural Chemistry in the Seventeenth Century," *Isis* 43 (1952): 12–36; J. E. McGuire, "Boyle's Conception of Nature," *Journal of the History of Ideas* 33 (1972): 523–42; and James R. Jacob's *Robert Boyle* (p. 204 below); and for philosophical surveys of matter theory and its connections with issues concerning the conditions of knowledge, see Peter Alexander, *Ideas, Qualities and Corpuscles: Locke and Boyle on the External World* (Cambridge: Cambridge University Press, 1985), and Maurice Mandelbaum, "Newton and Boyle and the Problem of 'Transdiction,'" in his *Philosophy, Science, and Sense Perception: Historical and Critical Studies* (Baltimore: Johns Hopkins University Press, 1966), 61–117. An important dissenting voice on the inanimate status of matter in the new philosophy is John Henry, for example, his "Occult Qualities and the Experimental Philosophy: Active Principles in Pre-Newtonian Matter Theory," *History of Science* 24 (1986): 335–81.

F. Medicine, Anatomy, and Physiology

As in the case of chemistry, traditional historiography reached no consensus about the "revolutionary" nature of the changes undergone by early modern medicine and allied practices. Vesalius's development of an observation-based human anatomy and Harvey's discovery of the circulation of the blood have typically been taken as paradigms of "revolutionary" achievements in these areas, while the thrust of recent historiography has, again, been to recognize "traditional" elements in these "new" accomplishments or to exempt these fields altogether from the "revolutionized" domain.

A traditional survey of anatomical belief and practice is F. J. Cole, *A History of Comparative Anatomy from Aristotle to the Eighteenth Century* (New York: Dover Books, 1975; orig. publ. 1949); for biology, see Eric Nordenskiöld, *The History of Biology: A Survey,* trans. Leonard Bucknall Eyre (New York: Tudor, 1946; orig. publ. 1920–24), esp. pt. 1, chaps. 11–13, and pt. 2, chaps. 1–4; for physiology, see Michael Foster, *Lectures on the History of Physiology during the Sixteenth, Seventeenth, and Eighteenth Centuries* (Cambridge: Cambridge University Press, 1901); Thomas S. Hall. *History of General Physiology,* 2 vols. (Chicago: University of Chicago Press, 1975; orig. publ. 1969), vol. 1, chaps. 11–24; and with special reference to seventeenth-century theories of respiration and nutrition, Everett Mendelsohn, *Heat and Life: The Development of the Theory of Animal Heat* (Cambridge: Harvard University Press, 1964), esp. chap. 3.

The history of medicine is an area where it is debatable whether recent historiography is really more sophisticated than that produced by some of the "founding figures," and an entry to the vast "traditional" literature on the history of early modern medicine can be had through, for example, Erwin H. Ackerknecht, *A Short History of Medicine,* rev. ed. (Baltimore: Johns Hopkins University Press, 1982; orig. publ. 1955), chaps. 9–10.

The definitive intellectual biography of Vesalius is C. D. O'Malley, *Andreas Vesalius of Brussels, 1514–1564* (Berkeley: University of California Press, 1964). On Harvey, there is now a rich traditional and revisionist literature: see, for example, Walter Pagel, *William Harvey's Biological Ideas: Selected Aspects and Historical Background* (Basel: S. Karger, 1967); Gweneth Whitteridge, *William Harvey and the Circulation of the Blood* (London: Macdonald, 1971); Jerome J. Bylebyl, "The Medical Side of Harvey's Discovery: The Normal and the Abnormal," in *William Harvey and His Age: The Medical and Social Context of the Discovery of the Circulation,* ed. Jerome J. Bylebyl, supplement to *Bulletin of the History of Medicine,* n.s., 2 (Baltimore: Johns Hopkins University Press, 1979), 28–102 (especially good for the practical medical context of Harvey's discovery); and Roger French, *William Harvey's Natural Philosophy* (Cambridge: Cambridge University Press, 1994). For subsequent English physiological work in Harvey's tradition, see Robert G. Frank Jr., *Harvey and the Oxford Physiologists: Scientific Ideas and Social Interaction* (Berkeley: University of California Press, 1980), which also contains abundant material on institutional aspects of science in England, and the concise historiographical survey by Andrew Wear, "The Heart and Blood from Vesalius to Harvey," in *Companion to the History of Modern Science* (p. 172 above), 568–82.

Despite the grip of mechanical conceptions on medical and physiological thought during the seventeenth century, historians have not in the main found reasons to celebrate a body of notable, still-recognized achievements flowing from mechanism. Indeed, the mechanization of medicine and physiology has more usually been identified as a "failed" aspiration, and other bases, including social and political considerations, have often been adduced to account for the appeal of mechanism in these areas. For the English setting see, for example, Theodore M. Brown, "The College of Physicians and the Acceptance of Iatro-mechanism in England, 1665–1695," *Bulletin of the History of Medicine* 44 (1970): 12–30; idem, "Physiology and the Mechanical Philosophy in Mid-Seventeenth-Century England," *Bulletin of the History of Medicine* 51 (1977): 25–54; Anita Guerrini, "James Keill, George Cheyne and Newtonian Physiology, 1690–1740," *Journal of the History of Biology* 18 (1985): 247–66; idem, "The Tory Newtonians: Gregory, Pitcairne and Their Circle," *Journal of British Studies* 25 (1986): 288–311; idem, "Archibald Pitcairne and Newtonian Medicine," *Medical History* 31 (1987): 70–83; Charles Webster, *The Great Instauration* (p. 182 above); idem, "William Harvey and the Crisis of Medicine in Jacobean England," in *William Harvey and His Age* (p. 184 above), 1–27; and Christopher Hill, "William Harvey and the Idea of Monarchy," in *The Intellectual Revolution of the Seventeenth Century,* ed. Charles Webster (London: Routledge and Kegan Paul, 1974), 160–81; see also Harold J. Cook, "The New Philosophy and Medicine in Seventeenth-Century England," in *Reappraisals of the Scientific Revolution* (p. 171 above), 397–436. For Cartesian mechanistic physiology and medicine, see the basic account in G. A. Lindeboom, *Descartes and Medicine* (Amsterdam: Rodopi, 1978); Thomas S. Hall, "The Physiology of Descartes," in his edition of Descartes's *Treatise of Man* (Cambridge: Harvard University Press, 1972), xxvi–xlviii; Leonora Cohen Rosenfield, *From Beast-Machine to Man-Machine: Animal Soul in French Letters from Descartes to La Mettrie,* new ed. (New York: Octagon Books, 1968; orig. publ. 1941); Richard B. Carter, *Descartes' Medical Philosophy: The Organic Solution to the Mind-Body Problem* (Baltimore: Johns Hopkins University Press, 1983); and, especially, Gaukroger's biography of Descartes (p. 209 below), 269–92. And for critical assessment of mechanical and nonmechanical frameworks in a specific biological domain, see Daniel Fouke, "Mechanical and 'Organical' Models in Seventeenth-Century Explanations of Biological Reproduction," *Science in Context* 3 (1989): 365–82.

A notable trend in recent historiography has been a move away from the study of medical theory toward an attempted reconstruction of the realities of medical practice, including an appreciation of the patient's point of view, though this idiom tends to depart from major concern with the Scientific Revolution. Among the representative texts of this newer focus see, for example, Lucinda McCray Beier, *Sufferers and Healers: The Experience of Illness in Seventeenth-Century England* (London: Routledge and Kegan Paul, 1987); various essays in Roger French and Andrew Wear, eds., *The Medical Revolution of the Seventeenth Century* (Cambridge: Cambridge University Press, 1989), and in Roy Porter, ed., *Patients and Practitioners: Lay Perceptions of Medicine in Pre-industrial Society* (Cambridge: Cambridge University Press, 1985). For lay female medical practice see, for example, Linda Pollock, *With Faith and Physic: The Life of a Tudor Gentlewoman, Lady Grace Mildmay, 1552–1620* (London: Collins and Brown, 1992), and for Boyle's extensive medical practice, see Barbara Beigun Kaplan, *"Divulging of Useful Truths in Physick": The Medical Agenda of Robert Boyle* (Baltimore: Johns Hopkins University Press, 1993). For an important episode in the debates over the natural versus supernatural character of medical curing, see Eamon Duffy, "Valentine Greatrakes, the Irish Stroker: Miracle, Science and Orthodoxy in Restoration England," *Studies in Church History* 17 (1981): 251–73; *Barbara Beigun Kaplan, "Greatrakes the Stroker: The Interpretations of His Contemporaries," *Isis* 73 (1982): 178–85; and *James R. Jacob, *Henry Stubbe, Radical Protestantism and the Early Enlightenment* (Cambridge: Cambridge University Press, 1983), chap. 3 and 164–74. The "weapon salve" of Sir Kenelm Digby (discussed in chapter 1 of this book) has not been subjected to much systematic modern historical treatment, but see Sir William Osler, *Sir Kenelm Digby's Powder of Sympathy: An Unfinished Essay by Sir William Osler* (Los Angeles: Plantin Press, 1972; orig. publ. 1900), and Betty Jo Teeter Dobbs, "Studies in the Natural Philosophy of Sir Kenelm Digby, Parts I–III," *Ambix* 18 (1971): 1–25; 20 (1973): 143–63; 21 (1974): 1–28.

G. Natural History and Related Practices

In seventeenth-century usage "natural history" referred to a register of "facts" about nature. Practitioners differed importantly about the relation of such a register to authentic "natural philosophy." The dominant tendency among English workers was to follow Bacon in attempting to reform natural history as a foundation for a reformed natural philosophy, whereas some English philosophers (e.g., Hobbes) and many Continentals (e.g., Descartes) reckoned that such a register—no matter how carefully it was assembled—could never *found* a systematic, secure, and certain philosophy of nature. Historians' engagements with the natural historical sciences have in the main followed seventeenth-century disagreements. For some, natural history does not belong to the mainstream of the Scientific Revolution, while others have drawn special attention to radical reforms in modes of observation and reporting that vastly extended the stock of empirical knowledge and that sought to distinguish more reliably between authentic and spurious observations of nature's existents. This contested division is currently one of the most basic in the historiography of the Scientific Revolution and even in construing the identity of that revolution (see also works cited in section 4A below).

For compact surveys of natural history in the seventeenth century, see Phillip R. Sloan, "Natural History, 1670–1802," in *Companion to the History of Modern Science* (p. 172 above), 295–313, and Joseph M. Levine, "Natural History and the History of the Scientific Revolution," *Clio* 13 (1983): 57–73; for sophisticated treatments of various aspects of early modern natural history, see Nicholas Jardine, James A. Secord, and Emma C. Spary, eds., *Cultures of Natural History* (Cambridge: Cambridge University Press, 1996), especially chapters by Ashworth, Cook, Findlen, Johns, Roche, and Whitaker); for natural history in relation to medicine, see Harold J. Cook, "The Cutting Edge of a Revolution? Medicine and Natural History Near the Shores of the North Sea," in *Renaissance and Revolution* (p. 174 above), 45–61; and for the changing cultural setting of natural history from the Renaissance to the seventeenth century, see William B. Ashworth Jr., "Natural History and the Emblematic World View," in *Reappraisals of the Scientific Revolution* (p. 171 above), 303–32. For a highly influential programmatic interpretation of Renaissance natural history as a search for "similitudes," see Michel Foucault, *The Order of Things: An Archaeology of the Human Sciences* (London: Tavistock, 1970; orig. publ. 1966), esp. chaps. 2–3.

An important and accessible study of European engagements with the New World, and its new stock of natural historical existents, is *Anthony Grafton, *New Worlds, Ancient Texts: The Power of Tradition and the Shock of Discovery* (Cambridge: Belknap Press of Harvard University Press, 1992); see also Stephen Greenblatt, *Marvelous Possessions: The Wonder of the New World* (Chicago: University of Chicago Press, 1991); *Steven Shapin, *A Social History of Truth* (see p. 192 below), esp. chap. 5 and chap. 6, 243–58; and Wilma George, "Source and Background to Discoveries of New Animals in the Sixteenth and Seventeenth Centuries," *History of Science* 18 (1980): 79–

104. For the practice of natural history in the American colonies, see, for example, Raymond Phineas Stearns, *Science in the British Colonies of America* (Urbana: University of Illinois Press, 1970).

A fine treatment of the purposes and practices of natural historical *collecting* in late Renaissance and Baroque Italy is *Paula Findlen, *Possessing Nature: Museums, Collecting, and Scientific Culture in Early Modern Italy* (Berkeley: University of California Press, 1994); see also Jay Tribby, "Body/Building: Living the Museum Life in Early Modern Europe," *Rhetorica* 10 (1992): 139–63; Krzysztof Pomian, *Collectors and Curiosities: Paris and Venice, 1500–1800,* trans. Elizabeth Wiles-Portier (Cambridge: Polity Press, 1990; orig. publ. 1987); Joseph M. Levine, *Dr. Woodward's Shield: History, Science, and Satire in Augustan England* (Berkeley: University of California Press, 1977); Stan A. E. Mendyk, *"Speculum Britanniae": Regional Study, Antiquarianism, and Science in Britain to 1700* (Toronto: University of Toronto Press, 1989); Thomas DaCosta Kaufmann, *The Mastery of Nature: Aspects of Art, Science, and Humanism in the Renaissance* (Princeton: Princeton University Press, 1993), chap. 7; Paolo Rossi, "Society, Culture and the Dissemination of Learning," in *Science, Culture and Popular Belief* (p. 180 above), 143–75 (esp. 162–72); Arthur MacGregor, *Sir Hans Sloane: Collector, Scientist, Antiquary, Founding Father of the British Museum* (London: British Museum Press, 1994); and several essays in Oliver Impey and Arthur MacGregor, eds., *The Origins of Museums: The Cabinet of Curiosities in Sixteenth- and Seventeenth-Century Europe* (Oxford: Clarendon Press, 1985). For the relation between humanistic scholarship and observation and reporting in early modern botany, see *Karen Meier Reeds, *Botany in Medieval and Renaissance Universities* (New York: Garland, 1991; orig. Harvard University Ph.D. diss., 1975).

Notable studies of seventeenth-century views of the particularity of natural historical experience and its place in natural philosophical practice include *Peter Dear, "*Totius in Verba:* Rhetoric and Authority in the Early Royal Society," *Isis* 76 (1985): 145–61; *Lorraine J. Daston, "The Factual Sensibility," *Isis* 79 (1988): 452-70 (an essay review of recent work on the culture of collecting); idem, "Marvelous Facts and Miraculous Evidence in Early Modern Europe," in *Questions of Evidence: Proof, Practice, and Persuasion across the Disciplines,* ed. James Chandler, Arnold I. Davidson, and Harry Harootunian (Chicago: University of Chicago Press, 1994; art. orig. publ. 1991), 243–74; idem, "Baconian Facts, Academic Civility, and the Prehistory of Objectivity," *Annals of Scholarship* 8 (1991): 337–63; Katharine Park and Lorraine J. Daston, "Unnatural Conceptions: The Study of Monsters in Sixteenth- and Seventeenth-Century France and England," *Past and Present* 92 (1981): 20–54; idem, *Wonders and the Order of Nature, 1150–1750* (New York, N.Y.: Zone Books, 1997), esp. pt. 2; and *Barbara J. Shapiro, *Probability and Certainty in Seventeenth-Century England: A Study of the Relationships between Natural Science, Religion, History, Law, and Literature* (Princeton: Princeton University Press, 1983), esp. chap. 4 (for the relation between human and natural historical practice). See also much work on the constitution and role of natural historical and experimental experience cited in section 4A below.

For aspects of geology, see Gordon L. Davies, *The Earth in Decay: A History of Geomorphology, 1578–1878* (London: Macdonald, 1969), chaps. 1–3; Martin J. S. Rudwick, *The Meaning of Fossils: Episodes in the History of Palaeontology* (Chicago: University of Chicago Press, 1985; orig. publ. 1972), chaps. 1–2; Roy Porter, *The Making of Geology: Earth Science in Britain, 1660–1815* (Cambridge: Cambridge University Press, 1977), chaps. 1–3; John C. Greene, *The Death of Adam: Evolution and Its Impact on Western Thought* (New York: Mentor Books, 1961; orig. publ. 1959), chaps. 1–3; Paolo Rossi, *The Dark Abyss of Time: The History of the Earth and the History of Nations from Hooke to Vico,* trans. Lydia G. Cochrane (Chicago: University of Chicago Press, 1984; orig. publ. 1979); Rachel Laudan, *From Mineralogy to Geology: The Foundations of a Science, 1650–1830* (Chicago: University of Chicago Press, 1987), chaps. 1–2; and Levine, *Dr. Woodward's Shield* (p. 188 above), chaps. 2–3. For geography see, for example, David N. Livingstone, *The Geographical Tradition: Episodes in the History of a Contested Enterprise* (Oxford: Basil Blackwell, 1993), chaps. 2–3, and Lesley B. Cormack, "'Good Fences Make Good Neighbors': Geography as Self-Definition in Early Modern England," *Isis* 82 (1991): 639–61. For early modern understandings of the distribution of plants and animals, see Janet Browne, *The Secular Ark: Studies in the History of Biogeography* (New Haven: Yale University Press, 1983), chap. 1; and for meteorology, see H. Frisinger, *The History of Meteorology to 1800* (New York: Science History Publications, 1977).

H. Sciences of the Human Mind, Human Nature, and Human Culture

Almost no historian of science has argued a case for early modern "revolution" in the practices now known as psychology or sociology, and accordingly the historical literature on these subjects is sparse. On the other hand, the considerable *problems* posed by mechanism for philosophies of knowledge, mind, and moral conduct are standard topics in the history of philosophy, though few practitioners in this idiom show a genuinely historical sensibility, preferring to argue with or to endorse the views of Descartes, Hobbes, Locke, and others rather than to interpret them as historically situated writings.

A fine starting point for psychology is Graham Richards, *Mental Machinery: The Origins and Consequences of Psychological Ideas. Part 1: 1600–1850* (Baltimore: Johns Hopkins University Press, 1992); for social sciences, see I. Bernard Cohen, *Interactions: Some Contacts between the Natural Sciences and the Social Sciences* (Cambridge: MIT Press, 1994); idem, "The Scientific Revolution and the Social Sciences," in *The Natural and the Social Sciences: Some Critical and Historical Perspectives,* ed. I. Bernard Cohen (Dordrecht: Kluwer, 1994), 153–203; and for anthropology, see Margaret T. Hodgen, *Early Anthropology in the Sixteenth and Seventeenth Centuries* (Philadelphia: University of Pennsylvania Press, 1964). For conceptions of human history, see Shapiro, *Probability and Certainty,* chap. 4; Rossi, *The Dark Abyss of Time* (both p. 189 above); and Joseph M. Levine, *Humanism and History: Origins of Modern English Historiography* (Ithaca: Cornell University Press, 1987). There are now some superb, and culturally resonant, accounts of the understanding and treatment of mental illness in the seventeenth century: see, for example, Michael MacDonald, *Mystical Bedlam: Madness, Anxiety, and Healing in Seventeenth-Century England* (Cambridge: Cambridge University Press, 1981), and Roy Porter, *Mind-Forg'd Manacles: A History of Madness in England from the Restoration to the Regency* (Cambridge: Harvard University Press, 1987).

An important treatment of conceptions of the person and the self in seventeenth-century philosophy is Charles Taylor's *Sources of the Self: The Making of Modern Identity* (Cambridge: Cambridge University Press, 1989), esp. chaps. 8–10 (for Descartes, Locke, and Montaigne), while Norbert Elias's *The Civilizing Process,* trans. Edmund Jephcott, 2 vols. (Oxford: Basil Blackwell, 1978, 1983; orig. publ. 1939, 1969) is a sociological study of changing early modern formations of the self that has been importantly drawn on by a number of recent historians of the Scientific Revolution. Special mention should be made of the sweeping survey of changing moral philosophical idioms by *Alisdair MacIntyre, *After Virtue: A Study in Moral Theory,* 2d ed. (Notre Dame: University of Notre Dame Press, 1984; orig. publ. 1981), which draws attention to the effects of the Scientific Revolution and the Enlightenment in dissolving the very idea of a "science" of moral conduct.

One of the most significant recent tendencies in the history of early modern science has been a recognition of close links between the practice of science and that of humanistic scholarship. The making of the new and the recovery of the old, tendencies once seen to be in opposition, are now increasingly recognized as belonging to the same enterprise. Here the leading scholar is Anthony Grafton: see especially his *New Worlds, Ancient Texts* (p. 187 above); *idem, *Defenders of the Text: The Traditions of Scholarship in an Age of Science, 1450–1800* (Cambridge: Harvard University Press, 1991), esp. chap. 7; idem, *Joseph Scaliger: A Study in the History of Classical Scholarship*, 2 vols. (Oxford: Clarendon Press, 1983–93); Anthony Grafton and Lisa Jardine, *From Humanism to Humanities: Education and the Liberal Arts in Fifteenth- and Sixteenth-Century Europe* (Cambridge: Harvard University Press, 1986); Anthony Grafton and Ann Blair, eds., *The Transmission of Culture in Early Modern Europe* (Philadelphia: University of Pennsylvania Press, 1990).

Other notable work on the relation between humanism and early modern science includes Barbara J. Shapiro, "Early Modern Intellectual Life: Humanism, Religion and Science in Seventeenth-Century England," *History of Science* 29 (1991): 45–71; Michael R. G. Spiller, *"Concerning Natural Experimental Philosophie": Meric Casaubon and the Royal Society* (The Hague: M. Nijhoff, 1980); Ann Blair, "Humanist Methods in Natural Philosophy: The Commonplace Book," *Journal of the History of Ideas* 53 (1992): 541–51; idem, "Tradition and Innovation in Early Modern Natural Philosophy: Jean Bodin and Jean-Cecile Frey," *Perspectives on Science* 2 (1994): 428–54; idem, *The Theater of Nature: Jean Bodin and Renaissance Science* (Princeton: Princeton University Press, 1997); Stephen Gaukroger, ed., *The Uses of Antiquity: The Scientific Revolution and the Classical Tradition* (Dordrecht: Kluwer, 1991); Lynn Sumida Joy, *Gassendi the Atomist: Advocate of History in an Age of Science* (Cambridge: Cambridge University Press, 1987); Jardine, *The Birth of History and Philosophy of Science* (p. 178 above); Dear, *Discipline and Experience* (p. 192 below), esp. chap. 4; Shapin, "'A Scholar and a Gentleman'" (p. 174 above); and Kaufmann, *The Mastery of Nature* (p. 188 above), esp. chaps. 5–6.

4. Topics and Themes

A. Experiment, Experience, and the Distribution of Knowledge

One of the characteristic marks of current historiography of early modern science is a heightened concern with the *practices* by which scientific knowledge was constituted, and this book is a robust expression of the value of such work for a historical understanding of what was both new and traditional about the Scientific Revolution. Naturally enough, my own work in this area has strongly influenced the overall conception of this book, most especially in chapters 2 and 3. *Steven Shapin and Simon Schaffer, *Leviathan and the Air-Pump: Hobbes, Boyle, and the Experimental Life* (Princeton: Princeton University Press, 1985), is an extended study of the social forms of knowledge making in English experimental philosophy, drawing out the general significance of the 1660s controversy between Boyle and Hobbes for an interpretation of the relations between social order and intellectual order. This work has been widely, and often quite critically, commented on by historians, philosophers, and sociologists interested in wider theoretical and methodological matters; see, for example, Bruno Latour, *We Have Never Been Modern,* trans. Catherine Porter (Cambridge: Harvard University Press, 1993; orig. publ. 1991), and Howard Margolis, *Paradigms and Barriers: How Habits of Mind Govern Scientific Beliefs* (Chicago: University of Chicago Press, 1993), chap. 11.

Related essays include *Steven Shapin, "Pump and Circumstance: Robert Boyle's Literary Technology," *Social Studies of Science* 14 (1984): 481–520; *idem, "The House of Experiment in Seventeenth-Century England," *Isis* 79 (1988): 373–404; idem, "'The Mind Is Its Own Place': Science and Solitude in Seventeenth-Century England," *Science in Context* 4 (1991): 191–218; *idem, "'A Scholar and a Gentleman'" (p. 174 above); and idem, "Who Was Robert Hooke?" in *Robert Hooke: New Studies* (p. 210 below), 253–85. More recently, I have stressed the significance of "gentlemanly" codes of conduct in the emerging practices through which early modern scientific practitioners constituted their stock of factual knowledge about the natural world: *Steven Shapin, *A Social History of Truth: Civility and Science in Seventeenth-Century England* (Chicago: University of Chicago Press, 1994).

My own work, while situated within a framework of potentially general interpretative interest, has concentrated on English materials, and the present book suggests that previously slighted aspects of seventeenth-century English practice can be seen as central to inquiries about what was "new" in the Scientific Revolution. Nevertheless, historical concern with the practices of constituting and warranting experience has also importantly treated Continental and, especially, Jesuit attitudes to experience. Here the outstanding work is by Peter Dear. His recent **Discipline and Experience: The Mathematical Way in the Scientific Revolution* (Chicago: University of Chicago Press, 1995) is a major contribution to understanding not only changing conceptions of experience but also their relation to the disciplines of mathematics and philosophy; see also Dear's "*Totius in Verba*" (p. 188 above) and his *"Miracles, Experiments, and the Ordinary Course of Nature," *Isis* 81 (1990): 663–83 (particularly for Pascal's Puy de Dôme experiment); and also Charles B. Schmitt's classic "Experience and Experiment: A Comparison of Zabarella's View with Galileo's in *De motu*," *Studies in the Renaissance* 16 (1969): 80–137. Further important writing on the constitution and reporting of experience includes Lorraine J. Daston's work cited above (pp. 188–89); Julian Martin, *Francis Bacon, the State, and the Reform of Natural Philosophy* (Cambridge: Cambridge University Press, 1992); Michael Aaron Dennis, "Graphic Understanding: Instruments and Interpretation in Robert Hooke's *Micrographia*," *Science in Context* 3 (1989): 309–64; Peter Dear, "Narratives, Anecdotes, and Experiments: Turning Experience into Science in the Seventeenth Century," in *The Literary Structure of Scientific Argument: Historical Studies*, ed. Peter Dear (Philadelphia: University of Pennsylvania Press, 1991), 135–63; Henry Krips, "Ideology, Rhetoric, and Boyle's *New Experiments*," *Science in Context* 7 (1994): 53–64; Christian Licoppe, *La formation de la pratique scientifique: Le discours de l'expérience en France et en Angleterre (1630–1820)* (Paris: Editions la Découverte, 1996); and Daniel Garber, "Experiment, Community, and the Constitution of Nature in the Seventeenth Century," *Perspectives on Science* 3 (1995): 173–201. One of the most sensitive and detailed studies of seventeenth-century experimental practice and of inference from experiment is *Simon Schaffer's essay on Newton's "crucial" prism experiments: "Glass Works: Newton's Prisms and the Uses of Experiment," in *The Uses of Experiment: Studies in the Natural Sciences*, ed. David Gooding, Trevor Pinch, and Simon Schaffer (Cambridge: Cambridge University Press, 1989), 67–104, on which episodes see also *Zev Bechler, "Newton's 1672 Optical Controversies: A Study in the Grammar of Scientific Dissent," in *The Interaction between Science and Philosophy*, ed. Yehuda Elkana (Atlantic Highlands, N.J.: Humanities Press, 1974), 115–42.

For the Italian setting, Paula Findlen's *Possessing Nature* (p. 188 above), her "Controlling the Experiment: Rhetoric, Court Patronage and the Experimental Method of Francesco Redi," *History of Science* 31 (1993): 35–64, and Jay Tribby's "Club Medici: Natural Experiment and the Imagineering of 'Tuscany,'" *Configurations* 2 (1994): 215–35, are relevant in connection with natural history and observational elements of experiment, while Biagioli's *Galileo, Courtier* and Winkler's and Van Helden's essays are important for observational astronomy in Italy, northern Europe, and other Continental settings (p. 178 above). For the assimilation of the microscope to seventeenth-century science, see *Catherine Wilson, *The Invisible World: Early Modern Philosophy and the Invention of the Microscope* (Princeton: Princeton University Press, 1995), and Barbara M. Stafford, *Body Criticism: Imaging the Unseen in Enlightenment Art and Medicine* (Cambridge: MIT Press, 1991), chap. 5; and for Leeuwenhoek and the practice of microscopic observation, see Clifford Dobell, *Antony van Leeuwenhoek and His "Little Animals": Being Some Account of the Father of Protozoology and Bacteriology and His Multifarious Discoveries in These Disciplines* (New York: Russell and Russell, 1958; orig. publ. 1932); Edward G. Ruestow, *The Microscope in the Dutch Republic: The Shaping of Discovery* (Cambridge: Cambridge University Press, 1996); and L. C. Palm and H. A. M. Snelders, eds., *Antoni van Leeuwenhoek, 1632–1723* (Amsterdam: Rodopi, 1982). *Paul Feyerabend's *Against Method: Outline of an Anarchistic Theory of Knowledge* (London: Verso, 1978; orig. publ. 1975), chaps. 9–10, sets out an important and provocative challenge to traditional understandings of Galileo's telescopic observations and how they were assessed; see also Vasco Ronchi, "The Influence of the Early Development of Optics on Science and Philosophy," in McMullin, ed., *Galileo: Man of Science* (p. 208 below), 195–206.

The introduction to the present book notes that the overwhelming majority of early modern Europeans did not participate in any type of organized science, or even in forms of literate knowledge, still less in the Scientific Revolution. Even so, the purposeful distinction between esoteric scientific knowledge and the beliefs of the "vulgar" common people was an absolutely fundamental concern of major practitioners from Galileo to Newton. Similarly, the proper cultural place of genuinely scientific knowledge, as between "private" and "secret" or "public" and "open," was intensely debated. Much of the literature on alchemy and astrology cited in sections 3C and 3E above deals with such concerns; see also William Eamon, *Science and the Secrets of Nature: Books of Secrets in Medieval and Early Modern Culture* (Princeton: Princeton University Press, 1994); J. V. Golinski, "A Noble Spectacle: Research on Phosphorus and the Public Cultures of Science in the Early Royal Society," *Isis* 80 (1989): 11–39; and Pamela O. Long, "The Openness of Knowledge: An Ideal and Its Context in Sixteenth-Century Writings on Mining and Metallurgy," *Technology and Culture* 32 (1991): 318–55.

For important studies of "low" knowledge in the early modern period and its placement vis-à-vis expert knowledge, see also Natalie Zemon Davis, *Society and Culture in Early Modern France* (Stanford: Stanford University

Press, 1975), chaps. 7–8; Carlo Ginzburg, *The Cheese and the Worms: The Cosmos of a Sixteenth-Century Miller,* trans. John and Anne C. Tedeschi (Harmondsworth: Penguin, 1982; orig. publ. 1976); idem, "The High and the Low: The Theme of Forbidden Knowledge in the Sixteenth and Seventeenth Centuries," in his *Clues, Myths, and the Historical Method,* trans. John and Anne C. Tedeschi (Baltimore: Johns Hopkins University Press, 1989; art. orig. publ. 1976), 60–76; *Christopher Hill, *The World Turned Upside Down: Radical Ideas during the English Revolution* (Harmondsworth: Penguin, 1975); Peter Burke, *Popular Culture in Early Modern Europe* (London: Temple Smith, 1978); various essays in Pumfrey et al., eds., *Science, Culture and Popular Belief* (p. 180 above); and Keith Thomas's *Religion and the Decline of Magic* and *Man and the Natural World* (both p. 177 above). See also Michael Heyd, "The New Experimental Philosophy: A Manifestation of Enthusiasm or an Antidote to It?" *Minerva* 25 (1987): 423–40; idem, *"Be Sober and Reasonable": The Critique of Enthusiasm in the Seventeenth and Early Eighteenth Centuries* (Leiden: E. J. Brill, 1995); Shapin and Schaffer, *Leviathan and the Air-Pump* (p. 192 above), chap. 7; and much work by James R. Jacob and Margaret C. Jacob cited below (pp. 204–5).

B. Science, Religion, Magic, and the Occult

In the late Victorian period it was common to write about the "warfare between science and religion" and to presume that these two bodies of culture must always have been in conflict. However, it has been a very long time since these attitudes have been held by historians of science. In one way or another the intimate connections between science and religion have been a leading concern of both "Great Tradition" and more recent historiography of the Scientific Revolution. In the late 1930s the American sociologist Robert K. Merton argued that the strand of English Protestantism known as Puritanism offered a congenial environment for the institutionalization of science in the seventeenth century: Merton, *Science, Economy and Society in Seventeenth-Century England* (New York: Harper, 1970; orig. publ. 1938). The historiographical controversies over the so-called Merton thesis have continued to the present day: see, for example, I. Bernard Cohen, ed., *Puritanism and the Rise of Modern Science: The Merton Thesis* (New Brunswick, N.J.: Rutgers University Press, 1990), and several essays in Charles Webster, ed., *The Intellectual Revolution of the Seventeenth Century* (p. 185 above). For many years historians of the "internalist" persuasion countered Merton's claims in much the same temper that was used to reject such Marxist historiography as Boris Hessen's (p. 173 above), although it is far from clear that the general form of Merton's highly qualified and cautiously phrased thesis was ever properly understood by the historians most critical of it. For a sketch of Merton's claim as sociological theory, see Steven Shapin, "Understanding the Merton Thesis," *Isis* 79 (1988): 594–605, and for representative criticism of the Merton thesis by a historian working in Koyré's idiom, see A. Rupert Hall, "Merton Revisited, or Science and Society in the Seventeenth Century," *History of Science* 2 (1963): 1–15.

For the English setting there is now a very large body of historical writing about the role of seventeenth-century science as "handmaid" to Protestant religion, with special reference to the strand of culture known as "natural theology." Among many examples, see Richard S. Westfall, *Science and Religion in Seventeenth-Century England* (New Haven: Yale University Press, 1958), and John Dillenberger, *Protestant Thought and Natural Science: A Historical Introduction* (Notre Dame: Notre Dame University Press, 1988; orig. publ. 1960). That basic cultural association between science and religion being accepted, attention has largely shifted to the precise forms of the relationship and to the consequences of particular religious commitments for the formation and evaluation of scientific beliefs. For views of miracles in English thought, see R. M. Burns, *The Great Debate on Miracles: From Joseph Glanvill to David Hume* (Lewisburg, Pa.: Bucknell University Press, 1981), and for contrasts between Protestant and Catholic frameworks for identifying miracles, see Peter Dear, "Miracles, Experiments, and the Ordinary Course of Nature" (p. 193 above). For English mechanical philosophy and the spirit world, see Simon Schaffer, "Godly Men and Mechanical Philosophers: Souls and Spirits in Restoration Natural Philosophy," *Science in Context* 1 (1987): 55–85, and Shapin and Schaffer, *Leviathan and the Air-Pump* (p. 192 above), chaps. 5 and 7; for Cartesianism and its problematic English acceptance by Christian philosophers see, for example, Alan Gabbey, "Philosophia Cartesiana Triumphata: Henry More (1646–1671)," in *Problems of Cartesianism,* ed. R. Davis et al. (Toronto: McGill-Queens University Press, 1982), 171–250; for Boyle and Newton, see many of the works cited in sections 5E and 5H below; also David Kubrin, "Newton and the Cyclical Cosmos: Providence and the Mechanical Philosophy," *Journal of the History of Ideas* 28 (1967): 325–46; McGuire, "Boyle's Conception of Nature"; and Shapin, "Of Gods and Kings" (both p. 183 above); Frank E. Manuel, *The Religion of Isaac Newton: The Fremantle Lectures 1973* (Oxford: Clarendon

Press, 1974); idem, *Isaac Newton Historian* (Cambridge: Belknap Press of Harvard University Press, 1963); and Neal C. Gillespie, "Natural Order, Natural Theology and Social Order: John Ray and the 'Newtonian Ideology,'" *Journal of the History of Biology* 20 (1987): 1–49. For early modern science and atheism, see Michael Hunter, "The Problem of 'Atheism' in Early Modern England," *Transactions of the Royal Historical Society,* 5th ser., 35 (1985): 135–57; idem, "Science and Heterodoxy: An Early Modern Problem Reconsidered," in *Reappraisals of the Scientific Revolution* (p. 171 above), 437–60; and Samuel I. Mintz, *The Hunting of Leviathan: Seventeenth-Century Reactions to the Materialism and Moral Philosophy of Thomas Hobbes* (Cambridge: Cambridge University Press, 1962).

Such is the current acceptance of the constructive significance of links between science and religion in the seventeenth century that few modern studies—of whatever setting—fail to address them in some way. Accordingly, much of the work cited throughout this bibliographic essay is pertinent in these connections. Surveys of science-religion links throughout early modern Europe include Reijer Hooykaas, *Religion and the Rise of Modern Science* (Edinburgh: Scottish Academic Press, 1972); the apologetically oriented Eugene M. Klaaren, *Religious Origins of Modern Science: Belief in Creation in Seventeenth-Century Thought* (Grand Rapids, Mich.: William B. Eerdmans, 1977); Amos Funkenstein, *Theology and the Scientific Imagination from the Middle Ages to the Seventeenth Century* (Princeton: Princeton University Press, 1986); various papers in *God and Nature* (p. 177 above); and the useful introductory survey by John Hedley Brooke, *Science and Religion: Some Historical Perspectives* (Cambridge: Cambridge University Press, 1991), esp. chaps. 1–4. For special attention to the theological setting of conceptions of natural law, see, for example, Francis Oakley, *Omnipotence, Covenant, and Order: An Excursion in the History of Ideas from Abelard to Leibniz* (Ithaca: Cornell University Press, 1984); Margaret Osler, *Divine Will and the Mechanical Philosophy: Gassendi and Descartes on Contingency and Necessity in the Created World* (Cambridge: Cambridge University Press, 1994); and for Catholic versus Protestant conceptions of how proper knowledge was to be secured, see William B. Ashworth Jr., "Light of Reason, Light of Nature: Catholic and Protestant Metaphors of Scientific Knowledge," *Science in Context* 3 (1989): 89–107. In these and many other connections see the important study of early modern skepticism in relation to the grounds of religious belief by *Richard H. Popkin, *The History of Scepticism from Erasmus to Spinoza,* rev. ed. (Berkeley: University of California Press, 1979; orig. publ. 1960).

The canonical site for interpretations of the relations between science and Catholic religion has been Galileo's trial by the Roman Church in 1633, but even here a more nuanced appreciation has emerged of the issues at stake (see, e.g., works by Redondi, Blackwell, Fantoli, and Feldhay in section 5A below). Some of the finest recent history of early modern science has, in fact, focused on the natural philosophical and mathematical work of specific Catholic orders, with particular attention having been paid to the Jesuits. Special circumstances affected the science of the Jesuits and other Catholic orders, and there were special constraints affecting the scientific work of lay practitioners living in Catholic settings, but there is no longer any sustainable and interesting sense in which it can be said that the Catholic Church was "antiscientific" or even unambiguously opposed to "the new science." For a concise introduction to the area, see William B. Ashworth Jr., "Catholicism and Early Modern Science," in *God and Nature* (p. 177 above), 136–66. For representative studies of Jesuit science, see Dear, *Discipline and Experience* (p. 192 above); idem, "The Church and the New Philosophy," in *Society, Culture and Popular Belief* (p. 180 above), 119–39; Lattis, *Between Copernicus and Galileo* (p. 177 above); Rivka Feldhay, "Knowledge and Salvation in Jesuit Culture," *Science in Context* 1 (1987): 195–213; idem, "Catholicism and the Emergence of Galilean Science: A Conflict between Science and Religion?" *Knowledge in Society* 7 (1988): 139–63; Rivka Feldhay and Michael Heyd, "The Discourse of Pious Science," *Science in Context* 3 (1989): 109–42; and Steven J. Harris, "Transposing the Merton Thesis: Apostolic Spirituality and the Establishment of the Jesuit Scientific Tradition," *Science in Context* 3 (1989): 29–65. And for aspects of Catholicism in the English scientific setting, see John Henry, "Atomism and Eschatology: Catholicism and Natural Philosophy in the Interregnum," *British Journal for the History of Science* 15 (1982): 211–40. An interesting perspective on Jewish reaction to new scientific thought is David B. Ruderman, *Jewish Thought and Scientific Discovery* (New Haven: Yale University Press, 1995).

If the general significance of science-religion links is now universally acknowledged in recent historiography, the related question of constructive connections between reformed science and various magical and mystical traditions still generates controversy. After all, much of the rhetoric of mechanical philosophers identified their new practice as a powerful solvent of the animistic and anthropocentric orientations that were often said to characterize "Renaissance naturalism" and "neo-Platonism" even more than the old Aristotelianism. And into the twentieth century commentators on the Scientific Revolution saw the mechanical philosophy as one of the basic causes of the "disenchantment of the world." Yet it is the legitimacy of that rhetoric that is contested in much recent work, some of which has already been cited in this section, as well as in sections treating historical studies of alchemy, astrology, and medicine.

An entry into this rich literature can be had through the survey by John Henry, "Magic and Science in the Sixteenth and Seventeenth Centuries," in *Companion to the History of Modern Science* (p. 172 above), 583–96, and the excellent short study by *Charles Webster, *From Paracelsus to Newton: Magic and the Making of Modern Science* (Cambridge: Cambridge University Press, 1982). A treasure trove of material on a wide range of magical traditions, still immensely useful despite its dated sensibilities, is Lynn Thorndike, *A History of Magic and Experimental Science,* 8 vols. (New York: Columbia University Press, 1923–58). Among the most influential individual studies of magic, mysticism, and science from the Renaissance to the seventeenth century are, for example, *J. E. McGuire and P. M. Rattansi, "Newton and the 'Pipes of Pan,'" *Notes and Records of the Royal Society of London* 21 (1966): 108–43; Frances A. Yates, *Giordano Bruno and the Hermetic Tradition* (Chicago: University of Chicago Press, 1964); and D. P. Walker, *Spiritual and Demonic Magic from Ficino to Campanella* (London: Warburg Institute, 1958).

Specific aspects of the place of the "occult," and of what was understood by the "occult," in the Scientific Revolution, are addressed by Simon Schaffer, "Occultism and Reason," in *Philosophy, Its History and Historiography,* ed. A. J. Holland (Dordrecht: D. Reidel, 1985), 117–43; Brian P. Copenhaver, "Natural Magic, Hermeticism, and Occultism in Early Modern Science," in *Reappraisals of the Scientific Revolution* (p. 171 above), 261–301; John Henry, "Occult Qualities and the Experimental Philosophy"; and Shapin, "Of Gods and Kings" (both p. 183 above); and in important essays by *Keith Hutchison, "What Happened to Occult Qualities in the Scientific Revolution?" *Isis* 73 (1982): 233–53, and idem, "Supernaturalism and the Mechanical Philosophy," *History of Science* 21 (1983): 297–333 (which argue a provocative case for the *enhanced* importance of the redefined occult and the supernatural in the mechanical worldview). A fine study of the changing treatment of magical animals is Brian P. Copenhaver, "A Tale of Two Fishes: Magical Objects from Antiquity through the Scientific Revolution," *Journal of the History of Ideas* 52 (1991): 373–98.

For qualified rejection of magic-science links see, among many examples, Mary Hesse, "Reasons and Evaluation in the History of Science," in *Changing Perspectives in the History of Science,* ed. Mikuláš Teich and Robert M. Young (Cambridge: Cambridge University Press, 1973), 127–47;

A. Rupert Hall, "Magic, Metaphysics and Mysticism in the Scientific Revolution," in *Reason, Experiment, and Mysticism in the Scientific Revolution,* ed. M. L. Righini Bonelli and William R. Shea (New York: Science History Publications, 1975), 275–82; Brian Vickers, Introduction to *Occult and Scientific Mentalities in the Renaissance,* ed. Brian Vickers (Cambridge: Cambridge University Press, 1984), 1–55; and for measured criticism of Frances Yates's views in relation to Copernicanism, see Robert S. Westman, "Magical Reform and Astronomical Reform: The Yates Thesis Reconsidered," in *Hermeticism and the Scientific Revolution: Papers Read at a Clark Library Seminar, March 9, 1974,* ed. Westman and J. E. McGuire (Los Angeles: William Andrews Clark Memorial Library, 1977), 1–91.

C. Social Forms, Relations, and Uses of Science

Time was when "the social dimensions of science" were treated as a special—usually marginal—factor in studies of early modern science. And, indeed, contrasts between "the life of the mind" and "the life in society" have been pervasive in Western culture since antiquity. Philosophers, religious thinkers, and "scientists" have routinely been characterized as leading lives quite disengaged from the mundane concerns of those who make things, love, war, and political order. One of the basic issues contested between the "externalist" and "internalist" historiographies noted in section 2 above and in the introduction to this book was whether early modern science could be adequately understood without reference to social and economic considerations. Yet as this book's concluding section argued, the contrast between the intellectual (or the natural) and the social is in part a cultural product of the Scientific Revolution. That opposition is what we need to understand, and it should not therefore be unreflectively used as a resource in historical inquiry. The production, maintenance, and transmission of science are undeniably social processes—whether or not features of the "wider society" are involved in interpreting any given part of science—and much of the recent work noted above treats "social aspects" not as a marginal "factor" but as constitutive of the very nature of science.

Nevertheless, there are senses in which studies of the social organization and social relations of science still recognizably form a distinct genre in writing about the Scientific Revolution. For example, there is a rich literature on the formation and functioning of the scientific societies that began to be an important feature of scientific work in the seventeenth century. A useful overview of European societies is James E. McClellan III, *Science Reorganized: Scientific Societies in the Eighteenth Century* (New York: Columbia University Press, 1985), esp. chaps. 1–2 (for seventeenth-century origins), and Martha Ornstein's *The Role of Scientific Societies in the Seventeenth Century* (Chicago: University of Chicago Press, 1928), though dated, is still useful for many national settings. For the Royal Society of London—perhaps the premier scientific organization of the period from 1660 to 1710—there is much classic work, including Sir Henry Lyons, *The Royal Society, 1660–1940: A History of Its Administration under Its Charters* (Cambridge: Cambridge University Press, 1944), chaps. 1–4; Dorothy Stimson, *Scientists and Amateurs: A History of the Royal Society* (New York: Henry Schuman, 1948); Sir Harold Hartley, ed., *The Royal Society: Its Origins and Founders* (London: Royal Society, 1960); and Margery Purver, *The Royal Society: Concept and Creation* (Cambridge: MIT Press, 1967). More recent studies centering on the Royal Society include K. Theodore Hoppen, "The Nature of the Early Royal Society," *British Journal for the History of Science* 9 (1976): 1–24, 243–73, and, especially, detailed studies by Michael Hunter: *Science and Society in Restoration England* (Cambridge: Cambridge University Press, 1981), *Establishing the New Science: The Experience of the Early Royal Society* (Woodbridge: Boydell Press, 1989), and *Science and the Shape of Orthodoxy: Intellectual Change in Late Seventeenth-Century Britain* (Woodbridge: Boydell Press, 1995), the first two of which have comprehensive bibliographic essays; see also Hunter's examination of the Royal Society's membership: *The Royal Society and Its Fellows, 1660–1700: The Morphology of an Early Scientific Institution,* 2d ed. (Oxford: Alden Press, 1994; orig. publ. 1982); and for percursor initiatives, see Mark Greenglass, Michael Leslie, and Timothy Raylor, eds., *Samuel Hartlib and Universal Reformation: Studies in Intellectual Communication* (Cambridge: Cambridge University Press, 1995). Recent studies of particular aspects of scientific work in the Royal Society include Marie Boas Hall, *Promoting Experimental Learning: Experiment and the Royal Society, 1660–1727* (Cambridge: Cambridge University Press, 1991); J. L. Heilbron, *Physics at the Royal Society during Newton's Presidency* (Los Angeles: William Andrews Clark Memorial Library, 1983); Robert C. Iliffe, "'In the Warehouse': Privacy, Property and Priority in the Early Royal Society," *History of Science* 30 (1992): 29–68; Shapin and Schaffer, *Leviathan and the Air-Pump;* and Shapin, *A Social History of Truth* (both p. 192 above), esp. chaps. 6 and 8.

For French societies, and especially for the Montmor Academy as a precursor to the Paris Academy of Sciences, a still very useful source is Harcourt

Brown, *Scientific Organizations in Seventeenth Century France (1620–1680)* (New York: Russell and Russell, 1967; orig. publ. 1934); for the Paris Academy itself, see Roger Hahn, *The Anatomy of a Scientific Institution: The Paris Academy of Sciences, 1666–1803* (Berkeley: University of California Press, 1971); Claire Salomon-Bayet, *L'institution de la science et l'expérience du vivant: Méthode et expérience à l'Académie royale des sciences, 1666–1793* (Paris: Flammarion, 1978); and Alice Stroup, *A Company of Scientists: Botany, Patronage, and Community at the Seventeenth-Century Parisian Royal Academy of Sciences* (Berkeley: University of California Press, 1990). For a representative study of a French provincial scientific society, see David S. Lux, *Patronage and Royal Science in Seventeenth-Century France: The Académie de Physique in Caen* (Ithaca: Cornell University Press, 1989); for a survey of French provincial academies, see Daniel Roche, *Le siècle des lumières en province: Académies et académiciens provinciaux, 1680–1789*, 2 vols. (Paris: Mouton, 1978); and for organizational innovations in the distribution of scientific information, see Howard M. Solomon, *Public Welfare, Science and Propaganda in Seventeenth-Century France: The Innovations of Théophraste Renaudot* (Princeton: Princeton University Press, 1972).

An important examination of a Florentine circle of experimentalists is W. E. Knowles Middleton, *The Experimenters: A Study of the Accademia del Cimento* (Baltimore: Johns Hopkins University Press, 1971), and for the social organization of science in Florence, Rome, and other Italian settings, see much material in Biagioli, *Galileo, Courtier* (p. 178 above); Findlen, *Possessing Nature;* Tribby, "Body/Building" (both p. 188 above); idem, Tribby, "Club Medici" (p. 188 above); and W. E. Knowles Middleton, "Science in Rome, 1675–1700, and the Accademia Fisicomathematica of Giovanni Giustino Ciampiani," *British Journal for the History of Science* 8 (1975): 138–54. For Ireland see K. Theodore Hoppen, *The Common Scientist in the Eighteenth Century: A Study of the Dublin Philosophical Society, 1683–1708* (London: Routledge and Kegan Paul, 1970). There is not nearly as much historical work on scientific societies in other European countries as there is for England, France, and Italy, but useful material in these connections can be found in various contributions to Porter and Teich, eds., *The Scientific Revolution in National Context* (p. 171 above). For the organizational and social forms of humanistic scholarship in Europe—some of which overlaps with natural scientific culture—from the late seventeenth to the mid-eighteenth century, see Anne Goldgar, *Impolite Learning: Conduct and Community in the Republic of Letters, 1680–1750* (New Haven: Yale University Press, 1995). And for treatments of the social forms in which scientific knowledge was made, without necessarily specific reference to formal organizations, see, for example, Owen Hannaway, "Laboratory Design and the Aim of Science: Andreas Libavius versus Tycho Brahe," *Isis* 77 (1986): 585–610; Mario Biagioli, "Scientific Revolution, Social Bricolage, and Etiquette," in *The Scientific Revolution in National Context* (p. 171 above), 11–54; as well as much work by Dear, Findlen, Shapin, and Tribby cited in sections 3G and 4A above.

The detailed investigation of forms of scientific patronage has been notable feature of much recent historiography of the Scientific Revolution. Representative studies of the patronage of Copernican and Galilean astronomy have already been cited above in section 3, including Biagioli's important *Galileo, Courtier,* Westfall's "Science and Patronage," and Westman's "Proof, Poetics, and Patronage." For the patronage of chemistry in Germany, see Smith, *The Business of Alchemy* (p. 182 above), chap. 2; for the patronage of mathematics in Restoration England, see Willmoth, *Sir Jonas Moore* (p. 180 above); for Robert Hooke and patronage in the same setting, see Hunter, *Establishing the New Science* (p. 201 above), chap. 9; for patronage and experimental life science in Italy, see Paula Findlen, "Controlling the Experiment: (p. 193 above); for courtly patronage in the Spanish setting, see David Goodman, "Philip II's Patronage of Science and Engineering," *British Journal for the History of Science* 16 (1983): 49–66; and, for the patronage of botany in France, see Stroup, *A Company of Scientists* (p. 202 above). A fine collection of essays on scientific patronage is Bruce T. Moran, ed., *Patronage and Institutions: Science, Technology, and Medicine at the European Court, 1500–1750* (Woodbridge: Boydell Press, 1991), especially papers by Eamon, Findlen, Moran, and Smith.

Much of the controversy surrounding Marxist history of early modern science mentioned in section 2 centered on questions of the relations between science and technology (or the economy generally). It was a dominant tendency in early Marxist work to identify important causal connections between the thematics, the dynamics, and (sometimes) the conceptual content of science, and this claim was also a major feature of the Merton thesis (p. 195 above). In response, "internalists," particularly those inspired by Koyré, vigorously denied any such causal influence on science from economic concerns. Classic representative Marxist assertions of economic impact on the growth of science include Hessen's "Social and Economic Roots of Newton's 'Principia'" and Zilsel's "The Sociological Roots of Science" (both p. 173 above), and representative systematic internalist ripostes are A. Rupert Hall, *Ballis-*

tics in the Seventeenth Century (Cambridge: Cambridge University Press, 1952), and idem, "The Scholar and the Craftsman in the Scientific Revolution" (p. 174 above).

The ideological charge of the debates over these issues that marked the period to the 1960s has now largely subsided, and more recent studies tend to adopt a more relaxed, matter-of-fact, and interpretatively heterogeneous attitude to science-technology relations in the early modern. Among many examples, see essays by Marie Boas Hall, A. Rupert Hall, Richard S. Westfall, and David W. Waters in *The Uses of Science in the Age of Newton,* ed. John G. Burke (Berkeley: University of California Press, 1983), all of which continue to take a broadly skeptical attitude. A fine and accessible affirmation of important science-technology links is *Paolo Rossi, *Philosophy, Technology, and the Arts in the Early Modern Era,* trans. Salvator Attanasio, ed. Benjamin Nelson (New York: Harper and Row, 1970; orig. publ. 1962), and for a sophisticated assertion of a positive case, focusing more on significant utilitarian intentions, attitudes, and legitimations than on concrete outcomes, see Larry Stewart, *The Rise of Public Science: Rhetoric, Technology, and Natural Philosophy in Newtonian Britain, 1660–1750* (Cambridge: Cambridge University Press, 1992). For a fascinating study of the development of fluid mechanics in relation to Italian practical problems of water management, see Cesare S. Maffioli, *Out of Galileo: The Science of Waters, 1628–1718* (Rotterdam: Erasmus, 1994), esp. pts. 3 and 4.

Perhaps the most contentious recent genre of historical work on early modern science focuses on its moral, political, and social uses. Although basic claims about constitutive relations between science, religion, and morality (noted above, p. 195–97) have achieved widespread acceptance, some historians have proffered quite specific arguments about the uses of science in supporting and subverting social and political order, and the case has even been advanced that such considerations of contextual social use must be appreciated if we are to understand the actual form, content, and modes of practice of a range of early modern sciences. A now somewhat dated review of work in this fast moving field—with special attention to the English setting—is Steven Shapin, "Social Uses of Science," in *The Ferment of Knowledge,* ed. Rousseau and Porter (p. 175 above), 93–139.

The preeminent historians in this idiom since the early 1970s have been James R. Jacob and Margaret C. Jacob. Most of their work has treated the use of natural knowledge as a political legitimating resource in seventeenth- and early eighteenth-century England: see, among many studies, *James R. Jacob, *Robert Boyle and the English Revolution: A Study in Social and Intellec-*

tual Change (New York: Burt Franklin, 1977); idem, "Boyle's Atomism and the Restoration Assault on Pagan Naturalism," *Social Studies of Science* 8 (1978): 211–33; idem, "Restoration Ideologies and the Royal Society," *History of Science* 18 (1980): 25–38; idem, *Henry Stubbe* (p. 186 above); idem, "The Political Economy of Science in Seventeenth-Century England," in *The Politics of Western Science, 1640–1990*, ed. Margaret C. Jacob (Atlantic Highlands, N.J.: Humanities Press, 1994), 19–46; *James R. Jacob and Margaret C. Jacob, "The Anglican Origins of Modern Science: The Metaphysical Foundations of the Whig Constitution," *Isis* 71 (1980): 251–67; *Margaret C. Jacob, *The Newtonians and the English Revolution, 1689–1720* (Ithaca: Cornell University Press, 1976); idem, *The Radical Enlightenment: Pantheists, Freemasons, and Republicans* (London: George Allen and Unwin, 1981); and her synthetic survey of both English and Continental developments, *The Cultural Meaning of the Scientific Revolution* (New York: McGraw-Hill, 1988). Important precursors to the perspective characterizing the Jacobs's work include Kubrin's "Newton and the Cyclical Cosmos" (p. 196 above) (see also idem, "Newton's Inside Out! Magic, Class Struggle, and the Rise of Mechanism in the West," in *The Analytic Spirit: Essays in the History of Science in Honor of Henry Guerlac*, ed. Harry Woolf [Ithaca: Cornell University Press, 1981], 96–121); the neglected classic, *Rudolph W. Meyer, *Leibnitz and the Seventeenth-Century Revolution*, trans. J. P. Stern (Chicago: Henry Regnery, 1952; orig. publ. 1948); and the body of Marxist scholarship mentioned in section 2.

The Jacobs's writings have been vigorously criticized on evidential grounds by historians of an "internalist" disposition, while they now seem to more sociologically inclined historians to be weakened, on the one hand, by excessive concern with individuals' motivations and shaky means of inferring such motivations and, on the other hand, by an apparent inability to connect considerations of social use to specific scientific concepts and to mundane aspects of scientific knowledge making. Nevertheless, their work has been both innovative and a spur to further new historical perspectives on the Scientific Revolution. Much of the recent scholarship noted in section 4A took some of its inspiration from the Jacobs's studies.

D. The Instruments of Science

One particular type of constitutive relationship between seventeenth-century science and technology was never contested in "Great Tradition" historiography and has in fact recently been the object of renewed detailed historical investigation—the use of purposefully designed instruments in making scientific knowledge. Some studies treating the role of the microscope, telescope, barometer, and mathematical instruments have been noted already (sections 3C, 3D, and 4A), and Bennett's "The Mechanics' Philosophy and the Mechanical Philosophy" (p. 174 above) is a superb overview of the connections between instruments and forms of knowledge. The historical literature on scientific instruments is vast, but some additional points of entry include: Derek J. de Solla Price, "The Manufacture of Scientific Instruments from *c.* 1500 to *c.* 1700," in *A History of Technology,* ed. Charles Singer et al. (London: Oxford University Press, 1957), 3:620–47; idem, "Philosophical Mechanism and Mechanical Philosophy: Some Notes towards a Philosophy of Scientific Instruments," *Annali dell'Istituto e Museo di Storia della Scienza di Firenze* 5 (1980): 75–85; Albert Van Helden, "The Birth of the Modern Scientific Instrument, 1550–1700," in *The Uses of Science in the Age of Newton,* ed. Burke (p. 204 above), 49–84; W. D. Hackmann, "Scientific Instruments: Models of Brass and Aids to Discovery," and J. A. Bennett, "A Viol of Water or a Wedge of Glass," both in *The Uses of Experiment,* ed. Gooding, Pinch, and Schaffer (p. 193 above), 31–65 and 105–14; W. E. Knowles Middleton, *A History of the Thermometer and Its Uses in Meteorology* (Baltimore: Johns Hopkins University Press, 1966); and Albert Van Helden, *The Invention of the Telescope,* Transactions of the American Philosophical Society 67 (4) (Philadelphia: American Philosophical Society, 1977).

A notable feature linking some recent history of early modern science to tendencies in the sociology of scientific knowledge is attention to the problematic character of instrumentally produced knowledge. Shapin and Schaffer's *Leviathan and the Air-Pump* (p. 192 above) contains much material on the difficulties of performing and replicating experiments with the air pump, and Shapin's *A Social History of Truth* (p. 192 above), chaps. 6 and 8, draws attention to technical and intellectual problems arising from the reporting of instrumentally mediated observations and from the work relations of the experimental laboratory; see also Schaffer, "Glass Works"; Dennis, "Graphic Understanding"; Wilson, *The Invisible World;* Feyerabend's account in *Against Method* of problems in the acceptance of Galileo's telescopic observations (all section 4A above); Ruestow, *The Microscope in the Dutch Republic* (p. 194); and Ian Hacking, *Representing and Intervening: Introductory Topics in the Philosophy of Natural Science* (Cambridge: Cambridge University Press, 1983), chap. 11 (on microscopes). These works also dwell on the interesting relationship between what was made visible through the use of scientific instruments and how those phenomena were communicated and made credible to others. It is in these connections that historical studies of modes of visual representation and, more generally, of the role of print technology have recently assumed central importance in the understanding of early modern science. For printing and science, see the seminal work of Elizabeth Eisenstein, *The Printing Press as an Agent of Change: Communications and Cultural Transformations in Early-Modern Europe,* 2 vols. (New York: Cambridge University Press, 1979); also Natalie Zemon Davis, *Society and Culture* (pp. 194–5 above), chap. 7; Adrian Johns, *The Nature of the Book: Knowledge and Print in Early Modern England* (Chicago: University of Chicago Press, 1998); Henry E. Lowood and Robin E. Rider, "Literary Technology and Typographic Culture: The Instrument of Print in Early Modern Culture," *Perspectives on Science* 2 (1994): 1–37; Eamon, *Science and the Secrets of Nature* (p. 194 above), esp. pt. 2; Paolo Rossi, "Science, Culture and the Dissemination of Knowledge," and Luce Giard, "Remapping Knowledge, Reshaping Institutions," both in *Science, Culture and Popular Belief* (p. 180 above), respectively 143–75 and 19–47 (esp. 25–32). For new modes of visual representation, see also Winkler and Van Helden, "Johannes Hevelius and the Visual Language of Astronomy" (p. 178 above); Svetlana Alpers's *The Art of Describing: Dutch Art in the Seventeenth Century* (Harmondsworth: Penguin, 1989; orig. publ. 1983), for conventions of realistic representation in art and science; and William M. Ivins Jr., *Prints and Visual Communication* (Cambridge: MIT Press, 1969), for the conventions of engraving. And for artistic conventions and representation in human anatomy, see Glenn Harcourt, "Andreas Vesalius and the Anatomy of Antique Sculpture," *Representations* 17 (1987): 28–61.

5. *Persons and Their Projects*

Much relevant historical work on the contributions of individual scientific practitioners and the projects associated with them has been cited already. (The indispensable biographical source here is *The Dictionary of Scientific Biography,* ed. Charles Coulston Gillispie, 18 vols. [New York: Scribner, 1970–90].) This section merely lists some further references for a few major figures treated in this book and in other accounts of the Scientific Revolution, concentrating on work of a biographical (or intellectual biographical) nature or otherwise tightly focused on an individual's career in science.

A. Galileo Galilei

Ludovico Geymonat, *Galileo Galilei: A Biography and Inquiry into His Philosophy of Science,* trans. Stillman Drake (New York: McGraw-Hill, 1965; orig. publ. 1957); Ernan McMullin, ed., *Galileo: Man of Science* (New York: Basic Books, 1967); William R. Shea, *Galileo's Intellectual Revolution* (London: Macmillan, 1972); Stillman Drake, *Galileo Studies: Personality, Tradition, and Revolution* (Ann Arbor: University of Michigan Press, 1970); idem, *Galileo at Work: His Scientific Biography* (Chicago: University of Chicago Press, 1978); William A. Wallace, *Galileo and His Sources: The Heritage of the Collegio Romano in Galileo's Science* (Princeton: Princeton University Press, 1984); idem, *Galileo's Logic of Discovery and Proof* (Dordrecht: Kluwer, 1992); Pietro Redondi, *Galileo Heretic,* trans. Raymond Rosenthal (Princeton: Princeton University Press, 1987; orig. publ. 1983); Richard J. Blackwell, *Galileo, Bellarmine, and the Bible* (Notre Dame: Notre Dame University Press, 1991); Joseph C. Pitt, *Galileo, Human Knowledge, and the Book of Nature: Method Replaces Metaphysics* (Dordrecht: Kluwer, 1992); Annibale Fantoli, *Galileo: For Copernicanism and for the Church,* trans. George V. Coyne (Vatican City: Vatican Observatory, 1994); and Rivka Feldhay, *Galileo and the Church: Political Inquisition or Critical Dialogue?* (Cambridge: Cambridge University Press, 1995).

B. Francis Bacon

Benjamin Farrington, *Francis Bacon, Philosopher of Industrial Science* (New York: Henry Schuman, 1949); idem, *The Philosophy of Francis Bacon* (Chicago: University of Chicago Press, 1964); Paolo Rossi, *Francis Bacon: From Magic to Science,* trans. Sacha Rabinovitch (Chicago: University of Chicago Press, 1968); Lisa Jardine, *Francis Bacon: Discovery and the Art of Discourse* (New York: Cambridge University Press, 1974); Peter Urbach, *Francis Bacon's Philosophy of Science: An Account and a Reappraisal* (La Salle, Ill.: Open Court, 1987); Antonio Pérez-Ramos, *Francis Bacon's Idea of Science and the Maker's Knowledge Tradition* (Oxford: Clarendon Press, 1988); Julian Martin, *Francis Bacon* (p. 193 above); Robert K. Faulkner, *Francis Bacon and the Project of Progress* (Lanham, Md.: Rowman and Littlefield, 1993); B. H. G. Wormald, *Francis Bacon: History, Politics, and Science, 1561–1626* (Cambridge: Cambridge University Press, 1993); John E. Leary Jr., *Francis Bacon and the Politics of Science* (Ames: Iowa State University Press, 1994); and

Markku Peltonen, ed., *The Cambridge Companion to Bacon* (Cambridge: Cambridge University Press, 1996).

C. Thomas Hobbes

Frithiof Brandt, *Thomas Hobbes' Mechanical Conception of Nature* (Copenhagen: Levin and Munksgaard, 1928); Arnold A. Rogow, *Thomas Hobbes: Radical in the Service of Reaction* (New York: W. W. Norton, 1986); and Tom Sorell, ed., *The Cambridge Companion to Hobbes* (Cambridge: Cambridge University Press, 1996).

D. René Descartes

Stephen Gaukroger, ed., *Descartes: Philosophy, Mathematics and Physics* (Brighton: Harvester Press, 1980); Marjorie Grene, *Descartes* (Minneapolis: University of Minnesota Press, 1985); idem, *Descartes among the Scholastics* (Milwaukee: Marquette University Press, 1991); William R. Shea, *The Magic of Numbers and Motion: The Scientific Career of René Descartes* (Canton, Mass.: Science History Publications, 1991); Daniel Garber, *Descartes' Metaphysical Physics* (Chicago: University of Chicago Press, 1992); John Cottingham, ed., *The Cambridge Companion to Descartes* (Cambridge: Cambridge University Press, 1992); Stephen Voss, ed., *Essays on the Philosophy and Science of René Descartes* (p. 172 above); *Stephen Gaukroger, *Descartes: An Intellectual Biography* (Oxford: Oxford University Press, 1995); and Roger Ariew and Marjorie Grene, eds., *Descartes and His Contemporaries: Meditations, Objections, and Replies* (Chicago: University of Chicago Press, 1995).

E. Robert Boyle

Louis Trenchard More, *The Life and Works of the Honourable Robert Boyle* (London: Oxford University Press, 1944); R. E. W. Maddison, *The Life of the Honourable Robert Boyle F.R.S.* (London: Taylor and Francis, 1969); James Jacob, *Robert Boyle* (p. 204 above); Jonathan Harwood, ed., *The Early Essays and Ethics of Robert Boyle* (Carbondale: Southern Illinois University Press, 1991); Steven Shapin, "Personal Change and Intellectual Biography: The Case of Robert Boyle," *British Journal for the History of Science* 26 (1993): 335–

45; Michael Hunter, ed., *Robert Boyle Reconsidered* (Cambridge: Cambridge University Press, 1994); and Rose-Mary Sargent, *The Diffident Naturalist: Robert Boyle and the Philosophy of Experiment* (Chicago: University of Chicago Press, 1995).

F. Robert Hooke

Margaret 'Espinasse, *Robert Hooke* (London: Heinemann, 1956); F. F. Centore, *Robert Hooke's Contributions to Mechanics: A Study in Seventeenth Century Natural Philosophy* (The Hague: M. Nijhoff, 1970); J. A. Bennett, "Robert Hooke as Mechanic and Natural Philosopher," *Notes and Records of the Royal Society* 35 (1980): 33–48; Michael Hunter and Simon Schaffer, eds., *Robert Hooke: New Studies* (Woodbridge: Boydell Press, 1989); Stephen Pumfrey, "Ideas above His Station: A Social Study of Hooke's Curatorship of Experiments," *History of Science* 29 (1991): 1–44; and Robert Iliffe, "Material Doubts: Hooke, Artisan Culture and the Exchange of Information in 1670s London," *British Journal for the History of Science* 28 (1995): 285–318.

G. Christiaan Huygens

Arthur Bell, *Christian Huygens and the Development of Science in the Seventeenth Century* (New York: Longmans Green, 1947); H. J. M. Bos, M. J. S. Rudwick, H. A. M. Snelders, and R. P. W. Visser, eds., *Studies on Christiaan Huygens* (Lisse: Swets and Zeitlinger, 1980); Aant Elzinga, *On a Research Program in Early Modern Physics, with Special Reference to the Work of Ch[ristiaan] Huygens* (Göteborg: Akademiförlaget, 1972); and Joella G. Yoder, *Unrolling Time: Christiaan Huygens and the Mathematization of Nature* (Cambridge: Cambridge University Press, 1988).

H. Isaac Newton

Louis Trenchard More, *Isaac Newton: A Biography* (New York: Charles Scribner's, 1934); Frank E. Manuel, *A Portrait of Isaac Newton* (Cambridge: Belknap Press of Harvard University Press, 1968); Richard S. Westfall, *Never at Rest: A Biography of Isaac Newton* (Cambridge: Cambridge University Press, 1980) and his recent abridgement published as *The Life of Isaac*

Newton (Cambridge: Cambridge University Press [Canto edition], 1994); Gale E. Christianson, *In the Presence of the Creator: Isaac Newton and His Times* (New York: Free Press, 1984); John Fauvel, Raymond Flood, Michael Shortland, and Robin Wilson, eds., *Let Newton Be!* (Oxford: Oxford University Press, 1988); and A. Rupert Hall, *Isaac Newton, Adventurer in Thought* (Oxford: Basil Blackwell, 1992).

Index